T0181926

# Tidy Finance with Python

This textbook shows how to bring theoretical concepts from finance and econometrics to the data. Focusing on coding and data analysis with Python, we show how to conduct research in empirical finance from scratch. We start by introducing the concepts of tidy data and coding principles using pandas, numpy, and plotnine. Code is provided to prepare common open-source and proprietary financial data sources (CRSP, Compustat, Mergent FISD, TRACE) and organize them in a database. We reuse these data in all the subsequent chapters, which we keep as self-contained as possible. The empirical applications range from key concepts of empirical asset pricing (beta estimation, portfolio sorts, performance analysis, Fama-French factors) to modeling and machine learning applications (fixed effects estimation, clustering standard errors, difference-in-difference estimators, ridge regression, Lasso, Elastic net, random forests, neural networks) and portfolio optimization techniques.

**Key Features:**

- Self-contained chapters on the most important applications and methodologies in finance, which can easily be used for the reader's research or as a reference for courses on empirical finance.
- Each chapter is reproducible in the sense that the reader can replicate every single figure, table, or number by simply copying and pasting the code we provide.
- A full-fledged introduction to machine learning with scikit-learn based on tidy principles to show how factor selection and option pricing can benefit from Machine Learning methods.
- We show how to retrieve and prepare the most important datasets financial economics: CRSP and Compustat, including detailed explanations of the most relevant data characteristics.
- Each chapter provides exercises based on established lectures and classes which are designed to help students to dig deeper. The exercises can be used for self-studying or as a source of inspiration for teaching exercises.

# Chapman & Hall/CRC
# The Python Series

## About the Series

Python has been ranked as the most popular programming language, and it is widely used in education and industry. This book series will offer a wide range of books on Python for students and professionals. Titles in the series will help users learn the language at an introductory and advanced level, and explore its many applications in data science, AI, and machine learning. Series titles can also be supplemented with Jupyter notebooks.

**Image Processing and Acquisition using Python, Second Edition**
*Ravishankar Chityala and Sridevi Pudipeddi*

**Python Packages**
*Tomas Beuzen and Tiffany-Anne Timbers*

**Statistics and Data Visualisation with Python**
*Jesús Rogel-Salazar*

**Introduction to Python for Humanists**
*William J.B. Mattingly*

**Python for Scientific Computation and Artificial Intelligence**
*Stephen Lynch*

**Learning Professional Python Volume 1: The Basics**
*Usharani Bhimavarapu and Jude D. Hemanth*

**Learning Professional Python Volume 2: Advanced**
*Usharani Bhimavarapu and Jude D. Hemanth*

**Learning Advanced Python from Open Source Projects**
*Rongpeng Li*

**Foundations of Data Science with Python**
*John Mark Shea*

**Data Mining with Python: Theory, Applications, and Case Studies**
*Di Wu*

**A Simple Introduction to Python**
*Stephen Lynch*

**Introduction to Python: with Applications in Optimization, Image and Video Processing, and Machine Learning**
*David Baez-Lopez and David Alfredo Báez Villegas*

For more information about this series please visit: https://www.crcpress.com/Chapman--HallCRC/book-series/PYTH

# Tidy Finance with Python

Christoph Scheuch
Stefan Voigt
Patrick Weiss
Christoph Frey

CRC Press
Taylor & Francis Group
Boca Raton London New York

CRC Press is an imprint of the
Taylor & Francis Group, an **Informa** business

A CHAPMAN & HALL BOOK

Designed cover image: © Christoph Frey, Christoph Scheuch, Stefan Voigt and Patrick Weiss

First edition published 2024
by CRC Press
2385 NW Executive Center Drive, Suite 320, Boca Raton FL 33431

and by CRC Press
4 Park Square, Milton Park, Abingdon, Oxon, OX14 4RN

*CRC Press is an imprint of Taylor & Francis Group, LLC*

© 2024 Christoph Frey, Christoph Scheuch, Stefan Voigt and Patrick Weiss

Reasonable efforts have been made to publish reliable data and information, but the author and publisher cannot assume responsibility for the validity of all materials or the consequences of their use. The authors and publishers have attempted to trace the copyright holders of all material reproduced in this publication and apologize to copyright holders if permission to publish in this form has not been obtained. If any copyright material has not been acknowledged please write and let us know so we may rectify in any future reprint.

Except as permitted under U.S. Copyright Law, no part of this book may be reprinted, reproduced, transmitted, or utilized in any form by any electronic, mechanical, or other means, now known or hereafter invented, including photocopying, microfilming, and recording, or in any information storage or retrieval system, without written permission from the publishers.

For permission to photocopy or use material electronically from this work, access www.copyright.com or contact the Copyright Clearance Center, Inc. (CCC), 222 Rosewood Drive, Danvers, MA 01923, 978-750-8400. For works that are not available on CCC please contact mpkbookspermissions@tandf.co.uk

*Trademark notice*: Product or corporate names may be trademarks or registered trademarks and are used only for identification and explanation without intent to infringe.

*Library of Congress Cataloging-in-Publication Data*
Names: Frey, Christoph (Writer on finance), author. | Scheuch, Christoph, author. | Voigt, Stefan (College teacher), author. | Weiss, Patrick, author.
Title: Tidy finance with Python / Christoph Frey, Christoph Scheuch, Stefan Voigt and Patrick Weiss.
Description: First edition. | Boca Raton, FL : CRC Press, 2024. | Includes bibliographical references and index.
Identifiers: LCCN 2023058911 (print) | LCCN 2023058912 (ebook) | ISBN 9781032684291 (hardback) | ISBN 9781032676418 (paperback) | ISBN 9781032684307 (ebook)
Subjects: LCSH: Finance--Data processing. | Econometrics. | Python (Computer program language)
Classification: LCC HG104 .F73 2024 (print) | LCC HG104 (ebook) | DDC 332.0285/6--dc23/eng/20240326
LC record available at https://lccn.loc.gov/202305891
LC ebook record available at https://lccn.loc.gov/2023058912

ISBN: 978-1-032-68429-1 (hbk)
ISBN: 978-1-032-67641-8 (pbk)
ISBN: 978-1-032-68430-7 (ebk)

DOI: 10.1201/9781032684307

Typeset in Latin Modern font
by KnowledgeWorks Global Ltd.

*Publisher's note:* This book has been prepared from camera-ready copy provided by the authors.

# Contents

# V  Portfolio Optimization                                                                    179

# Appendices                                                                                    209

# Preface

## Why Does This Book Exist?

Over our academic careers, we are continuously surprised by the lack of publicly available code for seminal papers or even textbooks in finance. This opaqueness and secrecy is particularly costly for young, aspiring financial economists. To tackle this issue, we started working on *Tidy Finance* to lift the curtain on reproducible finance. These efforts resulted in the book *Tidy Finance with R* (Scheuch et al., 2023), which provides a fully transparent code base in R for many common financial applications.

Since the book's publication, we received great feedback from students and teachers alike. However, one of the most common comments was that many interested coders are constrained and have to use Python in their institutions. We really love R for data analysis tasks, but we acknowledge the flexibility and popularity of Python. Hence, we decided to increase our team of authors with a Python expert and extend Tidy Finance to another programming language following the same tidy principles.

## Who Should Read This Book?

We write this book for three audiences:

- Students who want to acquire the basic tools required to conduct financial research ranging from the undergraduate to graduate level. The book's structure is simple enough such that the material is sufficient for self-study purposes.
- Instructors who look for materials to teach courses in empirical finance or financial economics. We choose a wide range of topics, from data handling and factor replication to portfolio allocation and option pricing, to offer something for every course and study focus. We provide plenty of examples and focus on intuitive explanations that can easily be adjusted or expanded. At the end of each chapter, we provide exercises that we hope inspire students to dig deeper.
- Practitioners like portfolio managers who like to validate and implement trading ideas or data analysts or statisticians who work with financial data and who need practical tools to succeed.

## What Will You Learn?

The book is divided into five parts:

- The first part helps you to set up your Python development environment and introduces you to essential programming concepts around which our approach to Tidy Finance revolves.
- The second part provides tools to organize your data and prepare the most common datasets used in financial research. Although many important data are behind paywalls, we start by describing different open-source data and how to download them. We then move on to prepare two of the most popular datasets in financial research: CRSP and Compustat. Then, we cover corporate bond data from TRACE. We reuse the data from these chapters in all subsequent chapters. The last chapter of this part contains an overview of common alternative data providers for which direct access via R packages exists.
- The third part deals with key concepts of empirical asset pricing, such as beta estimation, portfolio sorts, performance analysis, and asset pricing regressions.
- In the fourth part, we apply linear models to panel data and machine learning methods to problems in factor selection and option pricing.
- The last part provides approaches for portfolio optimization and backtesting procedures.

Each chapter is self-contained and can be read individually. Yet, the data chapters provide an important background necessary for data management in all other chapters.

## What Won't You Learn?

This book is about empirical work. We believe that our comparative advantage is to provide a thorough implementation of typical approaches such as portfolio sorts, backtesting procedures, regressions, machine learning methods, or other related topics in empirical finance. While we assume only basic knowledge of statistics and econometrics, we do not provide detailed treatments of the underlying theoretical models or methods applied in this book. Instead, you find references to the seminal academic work in journal articles or textbooks for more detailed treatments. Moreover, although we enrich our implementations by discussing the nitty-gritty choices you face while conducting empirical analyses, we refrain from deriving theoretical models or extensively discussing the statistical properties of well-established tools.

In addition, we also do not explain all the functionalities and details about the Python functions we use. We only delve into the empirical research focus and data transformation logic and want to refer attentive readers to consult the package documentations for more information. In other words, this is not a book to learn Python from scratch. It is a book on how to use Python as a tool to produce consistent and replicable empirical results.

That being said, our book is close in spirit to other books that provide fully reproducible code for financial applications. We view them as complementary to our work and want to highlight the differences:

- Hilpisch (2018) is a great introduction to the power of Python for financial applications. It does a great job explaining the basics of the Python language, its programming structure, and packages like `pandas`, `SciPy`, and `numpy` and uses these methods for actual applications. The book and a series of follow-up books[1] from the same author about financial data science, artificial intelligence, and algorithmic trading primarily target

---

[1]https://home.tpq.io/books/

practitioners and have a hands-on focus. Our book, in contrast, emphasizes reproducibility and starts with the applications right away to utilize Python as the tool to perform data transformations and statistical analysis. Hence, we clearly focus on state-of-the-art applications for academic research in finance. Thus, we fill a niche that allows aspiring researchers or instructors to rely on a well-designed code base.

- Furthermore, Weiming (2019) and Kelliher (2022) are comprehensive introductions to quantitative finance with a greater focus on option pricing, quantitative modeling for various markets besides equity, and algorithmic trading. Again, these books are primarily written for finance professionals to introduce Python or enhance their Python knowledge.
- Coqueret and Guida (2023) constitutes a great compendium to our book concerning applications related to return prediction and portfolio formation. The book primarily targets practitioners and has a hands-on focus. Our book, in contrast, relies on the typical databases used in financial research and focuses on the preparation of such datasets for academic applications. In addition, our chapter on machine learning focuses on factor selection instead of return prediction.

Although we emphasize the importance of reproducible workflow principles, we do not provide introductions to some of the core tools that we relied on to create and maintain this book:

- Version control systems such as Git[2] are vital in managing any programming project. Originally designed to organize the collaboration of software developers, even solo data analysts will benefit from adopting version control. Git also makes it simple to publicly share code and allow others to reproduce your findings. We refer to Bryan (2022) for a gentle introduction to the (sometimes painful) life with Git.
- Good communication of results is a key ingredient to reproducible and transparent research. To compile this book, we heavily draw on a suite of fantastic open-source tools. First, Kibirige (2023b) provides a highly customizable yet easy-to-use system for creating data visualizations based on the Grammar of Graphics (Wilkinson, 2012). Second, in our daily work and to compile this book, we used Quarto, an open-source scientific and technical publishing system described in Allaire et al. (2023). Markdown documents are fully reproducible and support static and dynamic output formats. We do not provide introductions to these tools, as the resources above already provide easily accessible tutorials.
- Good writing is also important for the presentation of findings. We neither claim to be experts in this domain nor do we try to sound particularly academic. On the contrary, we deliberately use a more colloquial language to describe all the methods and results presented in this book in order to allow our readers to relate more easily to the rather technical content. For those who desire more guidance with respect to formal academic writing for financial economics, we recommend Kiesling (2003), Cochrane (2005), and Jacobsen (2014), who all provide essential tips (condensed to a few pages).

## Why Tidy?

As you start working with data, you quickly realize that you spend a lot of time reading, cleaning, and transforming your data. In fact, it is often said that more than 80 percent of

---

[2]https://git-scm.com/

data analysis is spent on preparing data. By *tidying data*, we want to structure datasets to facilitate further analyses. As Wickham (2014) puts it:

> [T]idy datasets are all alike, but every messy dataset is messy in its own way. Tidy datasets provide a standardized way to link the structure of a dataset (its physical layout) with its semantics (its meaning).

In its essence, tidy data follows these three principles:

1. Every column is a variable.
2. Every row is an observation.
3. Every cell is a single value.

Throughout this book, we try to follow these principles as best as possible. If you want to learn more about tidy data principles in an informal manner, we refer you to this vignette[3] as part of Wickham and Girlich (2022).

In addition to the data layer, there are also tidy coding principles outlined in the tidy tools manifesto[4] that we try to follow:

1. Reuse existing data structures.
2. Compose simple functions.
3. Embrace functional programming.
4. Design for humans.

---

## Why Python?

Python (Python Software Foundation, 2023) is an open-source, general-purpose, high-level programming language widely used across various industries. Python is prevalent for data science according to the Python Developers Survey (Foundation and JetBrains, 2022), particularly for financial applications. Similar to R, the Python community values readable and straightforward code. Thus, Python is an excellent choice for first-time programmers. At the same time, experienced researchers in financial economics and analysts benefit from the wide range of possibilities to express complex ideas with concise and understandable code. Some of the highlights of Python include:

- Open-source: Python uses a source license, making it usable and distributable for academic and commercial use.
- Flexibility: Python's extensive ecosystem of standard libraries and community-contributed modules allows for all kinds of unique projects. It seamlessly integrates various data sources and APIs, facilitating efficient data retrieval and processing.
- Versatility: Python is a cross-platform, multipurpose language that can be used to write fast low-level executable code, large applications, and even graphical user interfaces (GUI).
- Speed: Python is fast. In addition, parallelization is straightforward to implement in order to tackle big data problems without hassle.
- Robustness: Python provides robust tools for data manipulation, analysis, and visualization, which are crucial components in finance research.

---

[3]https://cran.r-project.org/web/packages/tidyr/vignettes/tidy-data.html
[4]https://tidyverse.tidyverse.org/articles/manifesto.html

- Importance: The language's active community support and continuous development ensure access to cutting-edge technologies and methodologies. Learning Python enhances one's ability to conduct sophisticated financial analysis, making it a valuable skill for professionals across diverse fields.

The so-called *Zen of Python* by Tim Peters summarizes its major syntax guidelines for structured, tidy, and human-readable code. It is easily accessible in every Python environment through:

```
import this
```

```
The Zen of Python, by Tim Peters

Beautiful is better than ugly.
Explicit is better than implicit.
Simple is better than complex.
Complex is better than complicated.
Flat is better than nested.
Sparse is better than dense.
Readability counts.
Special cases aren't special enough to break the rules.
Although practicality beats purity.
Errors should never pass silently.
Unless explicitly silenced.
In the face of ambiguity, refuse the temptation to guess.
There should be one-- and preferably only one --obvious way to do it.
Although that way may not be obvious at first unless you're Dutch.
Now is better than never.
Although never is often better than *right* now.
If the implementation is hard to explain, it's a bad idea.
If the implementation is easy to explain, it may be a good idea.
Namespaces are one honking great idea -- let's do more of those!
```

Python comes in many flavors, and endless external packages extend the possibilities for conducting financial research. Any code we provide echoes some arguably subjective decisions we have taken to comply with our idea of what Tidy Finance comprises: Code should not simply yield the correct output but should be easy to read. Therefore, we advocate using chaining, which is the practice of calling multiple methods in a sequence, each operating on the result of the previous step.

Further, the entire book rests on tidy data, which we handle with a small set of powerful packages proven effective: `pandas` and `numpy`. Regarding visualization (which we deem highly relevant to provide a fundamentally human-centered experience), we follow the Grammars of Graphics' philosophical framework (Wilkinson, 2012), which has been carefully implemented using `plotnine`.

Arguably, neither chaining commands nor using the Grammar of Graphics can be considered mainstream within the Python ecosystem for financial research (yet). We believe in the value of the workflows we teach and practice on a daily basis. Therefore, we also believe that adopting such coding principles will dramatically increase in the near future. For more information on why Python is great, we refer to Hilpisch (2018).

## License

This book is licensed to you under Creative Commons Attribution-NonCommercial-ShareAlike 4.0 International CC BY-NC-SA 4.0. The code samples in this book are licensed under Creative Commons CC0 1.0 Universal (CC0 1.0), i.e., public domain. You can cite the Python project as follows:

Scheuch, C., Voigt, S., Weiss, P., & Frey, C. (2024). *Tidy Finance with Python* (1st ed.). Chapman and Hall/CRC. https://www.tidy-finance.org.

You can also use the following BibTeX entry:

```
@book{Scheuch2024,
  title = {Tidy Finance with Python (1st ed.)},
  author = {and Scheuch, Christoph and Voigt, Stefan and Weiss, Patrick
          and Frey, Christoph},
  year = {2024},
  publisher = {Chapman and Hall/CRC},
  edition = {1st},
  url = {https://tidy-finance.org/python}
}
```

## Future Updates and Changes

This book represents a snapshot of research practices and available data at a particular time. However, time does not stop. As you read this text, there is new data, packages used here have changed, and research practices might be updated. We as authors of Tidy Finance are committed to staying up-to-date and keeping up with the newest developments. Therefore, you can expect updates to Tidy Finance on a continuous basis. The best way for you to monitor the ongoing developments, is to check our online Changelog[5] frequently.

---

[5] https://tidy-finance.org/python/changelog.html

# Author Biographies

**Christoph Scheuch** is the Head of Artificial Intelligence at the social trading platform wikifolio.com. He is responsible for researching, designing, and prototyping of cutting-edge AI-driven products using R and Python. Before his focus on AI, he was responsible for product management and business intelligence at wikifolio.com and an external lecturer at the Vienna University of Economics and Business, Austria, where he taught finance students how to manage empirical projects.

**Stefan Voigt** is an Assistant Professor of Finance at the Department of Economics at the University in Copenhagen, Denmark and a research fellow at the Danish Finance Institute. His research focuses on blockchain technology, high-frequency trading, and financial econometrics. Stefan's research has been published in the leading finance and econometrics journals and he received the Danish Finance Institute Teaching Award 2022 for his courses for students and practitioners on empirical finance based on Tidy Finance.

**Patrick Weiss** is an Assistant Professor of Finance at Reykjavik University, Iceland, and an external lecturer at the Vienna University of Economics and Business, Austria. His research activity centers around the intersection of empirical asset pricing and corporate finance, with his research appearing in leading journals in financial economics. Patrick is especially passionate about empirical asset pricing and strives to understand the impact of methodological uncertainty on research outcomes.

**Christoph Frey** is a Quantitative Researcher and Portfolio Manager at a family office in Hamburg, Germany, and a Research Fellow at the Centre for Financial Econometrics, Asset Markets and Macroeconomic Policy at Lancaster University, UK. Prior to this, he was the leading quantitative researcher for systematic multi-asset strategies at Berenberg Bank and worked as an Assistant Professor at the Erasmus Universiteit Rotterdam, Netherlands. Christoph published research on Bayesian Econometrics and specializes in financial econometrics and portfolio optimization problems.

# Part I

# Getting Started

# 1

# Setting Up Your Environment

We aim to lower the bar for starting empirical research in financial economics. We want to make using Python easy for you. However, given that Tidy Finance is a platform that supports multiple programming languages, we also consider the possibility that you are unfamiliar with Python. Maybe you are transitioning from R to Python, i.e., following the journey of Tidy Finance, which started in R. Hence, we provide you with a simple guide to get started with Python. If you have not used Python before, you will be able to use it after reading this chapter.

## 1.1 Python Environment

A Python environment is a self-contained directory or folder containing a specific version of the Python installation with a set of packages and dependencies. In order to isolate and manage the specific dependencies of the Tidy Finance with Python project, a so-called *virtual environment* is a reliable way to ensure that it will work consistently and reliably on different systems and over time.

There are many ways to install a Python version and environments on your system. We present two ways that we found most convenient to write this book and maintain our website: (i) Installation via Anaconda along with using Python in Spyder and (ii) installation via RStudio.

## 1.2 Installation via Anaconda

First, we need to install Python itself via Anaconda. You can download the latest version of Anaconda from the official Anaconda website.[1] After downloading the respective version for your operating system, follow the installation instructions.

Second, we now describe how to set up a Python virtual environment specific to Tidy Finance on your local system. This book uses Python version 3.10.11 to set up the environment for both Windows and Mac. As we write this book, it is not the latest version of Python. The reason for this is that we wanted (i) a stable code base and (ii) the content of the book to be usable for all kinds of users, especially for those who might rely on corporate version controls and are not able to install new Python distributions.

---

[1] https://www.anaconda.com/products/individual

For the installation, we use the Anaconda Python distribution you downloaded in the step before.[2] Additionally, you need the packages listed in the provided `requirements.txt`-file[3] in a dedicated folder for the project. You can find the detailed list of packages in Appendix A.

We recommend you start with the package installation right away. After you have prepared your system, you can open the Anaconda prompt and install your virtual environment with the following commands:

1. `conda create -p C:\Apps\Anaconda3\envs\tidy_finance_environment python==3.10.11` (Confirm with y)
2. `conda activate C:\Apps\Anaconda3\envs\tidy_finance_environment`
3. `pip install -r "<Tidy-Finance-with-Python Folder>\requirements.txt"`

All other packages found with the command pip list are installed automatically as dependencies with the required packages in the file `requirements.txt`. Note that we make reference to two distinct folders. The first one, `C:\Apps\Anaconda3\envs\tidy_finance_environment` refers to the location of your Python environment used for Tidy Finance. Apart from that, you should store your data, program codes, and scripts in another location: Your Tidy Finance working folder.

Now, you are basically ready to go. However, you will now need a Python integrated development environment (IDE) to make your coding experience pleasant.

## 1.3 Python IDE

If you are new to coding, you will not have an IDE for Python. We recommend using Spyder if you plan to code only in Python as it comes with Anaconda. If you don't use Anaconda, you can download the software for your operating system from the official website.[4] Then, follow the installation instructions. To add the previously created virtual environment to Spyder, Go to Tools → Preferences → Python Interpreter → "Use the following interpreter" and add `C:\Apps\Anaconda3\envs\tidy_finance_environment\python.exe`.

Another increasingly popular code editor for data analysis is Visual Studio Code (VS Code), as it supports a variety of programming languages, including Python and R. We refer to this tutorial[5] if you want to get started with VS Code. There are many more ways to set up a Python IDE, so we refer to this page[6] in the Python wiki for more inspiration.

If you also plan to try R, you should get a multipurpose tool: RStudio. You can get your RStudio version from Posit[7] (i.e., the company that created RStudio, which was previously called RStudio itself). When you follow the instructions, you will see that Posit asks you to install R; you need to do so to make RStudio feasible for Python. Then, select the virtual

---

[2]Note that you can also install a newer version of Python. We only require the environment set up in the previous step to use Python version 3.10.11. The neat aspect is Python's capability to accommodate version control in this respect.

[3]https://github.com/tidy-finance/website/blob/main/requirements.txt

[4]https://www.spyder-ide.org/

[5]https://code.visualstudio.com/docs/python/python-tutorial

[6]https://wiki.python.org/moin/IntegratedDevelopmentEnvironments

[7]https://posit.co/download/rstudio-desktop/

environment in RStudio. Alternatively, you can also start with the installation guide starting from RStudio, which we present below.

---

## 1.4  Installation via RStudio

You can also install Python and set up your environment directly from RStudio. This approach has the big advantage that you can switch between R and Python code smoothly. We believe that being able to switch between different programming languages is a tremendously valuable skill, so we set up a repository containing all the files that you need to achieve this goal: Tidy Finance Environment.[8] To set up this environment locally, follow these steps:

1. Install R[9] and RStudio.[10]
2. Clone the Tidy Finance Environment[11] repository directly in RStudio by clicking on **File/New Project/** and selecting **Version Control**. Then, click **Git** and provide the repository address https://github.com/tidy-finance/environment. RStudio will then automatically open the project in the new environment.
3. Install the **reticulate** R package: `install.packages("reticulate")`.
4. Use **reticulate** to install Python: `reticulate::install_python(version = "3.10.11", force = TRUE)`.
5. Tell **renv** to use Python: `renv::use_python("PATH")`.
   a. "PATH" on Mac: `"~/.pyenv/versions/3.10.11/bin/python"`.
   b. "PATH" on Windows: `"C:/Users/<User>/AppData/Local/r-reticulate/r-reticulate/pyenv/pyenv-win/versions/3.10.11/python.exe"` where `<User>` is your username.
6. Tell **renv** to install all required packages: `renv::restore()`.

Now you are ready to execute all code that you can find in this book or its sibling *Tidy Finance with R*.[12]

---

[8] https://github.com/tidy-finance/environment
[9] https://cran.r-project.org/
[10] https://posit.co/download/rstudio-desktop/
[11] https://github.com/tidy-finance/environment
[12] https://www.tidy-finance.org/r/index.html

# 2

## Introduction to Tidy Finance

The main aim of this chapter is to familiarize yourself with **pandas** (Wes McKinney, 2010) and **numpy** (Harris et al., 2020), the main workhorses for data analysis in Python. We start by downloading and visualizing stock data from Yahoo!Finance. Then, we move to a simple portfolio choice problem and construct the efficient frontier. These examples introduce you to our approach of *Tidy Finance*.

### 2.1 Working with Stock Market Data

At the start of each session, we load the required Python packages. Throughout the entire book, we always use **pandas** and **numpy** to perform a number of data manipulations. In this chapter, we also load the convenient **yfinance** (Aroussi, 2023) package to download price data.

```
import pandas as pd
import numpy as np
import yfinance as yf
```

Note that `import pandas as pd` implies that we can call all pandas functions later with a simple `pd.function()`. Instead, utilizing `from pandas import *` is generally discouraged, as it leads to namespace pollution. This statement imports all functions and classes from **pandas** into your current namespace, potentially causing conflicts with functions you define or those from other imported libraries. Using the **pd** abbreviation is a very convenient way to prevent this.

We first download daily prices for one stock symbol, e.g., the Apple stock, *AAPL*, directly from the data provider Yahoo!Finance. To download the data, you can use the function `yf.download()`. The data from Yahoo!Finance comes as a dataframe, a two-dimensional, tabular data structure in which each row is indexed, and each column has a name. After the download, we apply a set of functions directly on the dataframe. First, we put the date index into a separate column. Second, we add the column **symbol** that stores the ticker information, and finally, we rename all columns to lowercase names. dataframes allow for *chaining* all these operations sequentially through using ..

```
prices = (yf.download(
    tickers="AAPL",
    start="2000-01-01",
    end="2022-12-31",
    progress=False
  )
  .reset_index()
```

```
  .assign(symbol="AAPL")
  .rename(columns={
    "Date": "date",
    "Open": "open",
    "High": "high",
    "Low": "low",
    "Close": "close",
    "Adj Close": "adjusted",
    "Volume": "volume"}
  )
)
prices.head().round(3)
```

|   | date | open | high | low | close | adjusted | volume | symbol |
|---|------|------|------|-----|-------|----------|--------|--------|
| 0 | 2000-01-03 | 0.936 | 1.004 | 0.908 | 0.999 | 0.846 | 535796800 | AAPL |
| 1 | 2000-01-04 | 0.967 | 0.988 | 0.903 | 0.915 | 0.775 | 512377600 | AAPL |
| 2 | 2000-01-05 | 0.926 | 0.987 | 0.920 | 0.929 | 0.786 | 778321600 | AAPL |
| 3 | 2000-01-06 | 0.948 | 0.955 | 0.848 | 0.848 | 0.718 | 767972800 | AAPL |
| 4 | 2000-01-07 | 0.862 | 0.902 | 0.853 | 0.888 | 0.752 | 460734400 | AAPL |

`yf.download()` downloads stock market data from Yahoo!Finance. The above code chunk returns a dataframe with eight quite self-explanatory columns: `date`, the market prices at the `open`, `high`, `low`, and `close`, the `adjusted` price in USD, the daily `volume` (in the number of traded shares), and the `symbol`. The adjusted prices are corrected for anything that might affect the stock price after the market closes, e.g., stock splits and dividends. These actions affect the quoted prices but have no direct impact on the investors who hold the stock. Therefore, we often rely on adjusted prices when it comes to analyzing the returns an investor would have earned by holding the stock continuously.

Next, we use the `plotnine` (Kibirige, 2023b) package to visualize the time series of adjusted prices in Figure 2.1. This package takes care of visualization tasks based on the principles of the Grammar of Graphics (Wilkinson, 2012). Note that generally, we do not recommend using the * import style. However, we use it here only for the plotting functions, which are distinct to `plotnine` and have very plotting-related names. So, the risk of misuse through a polluted namespace is marginal.

```
from plotnine import *
```

Creating figures becomes very intuitive with the Grammar of Graphics, as the following code chunk demonstrates.

```
prices_figure = (
  ggplot(prices,
         aes(y="adjusted", x="date")) +
  geom_line() +
  labs(x="", y="",
       title="Apple stock prices from 2000 to 2022")
)
prices_figure.draw()
```

Instead of analyzing prices, we compute daily returns defined as

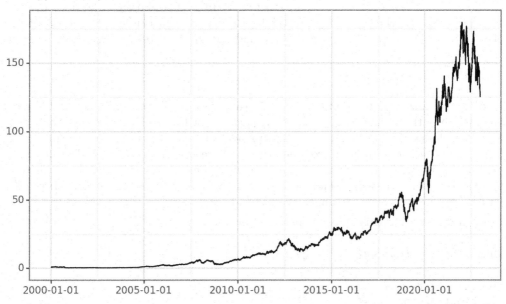

Apple stock prices from 2000 to 2022

Figure 2.1: The figure shows Apple stock prices between the beginning of 2000 and the end of 2022. Prices are in USD, adjusted for dividend payments and stock splits.

$$r_t = p_t/p_{t-1} - 1, \tag{2.1}$$

where $p_t$ is the adjusted price on day $t$. In that context, the function pct_change() is helpful because it computes this percentage change.

```
returns = (prices
  .sort_values("date")
  .assign(ret=lambda x: x["adjusted"].pct_change())
  .get(["symbol", "date", "ret"])
)
```

|   | symbol | date       | ret    |
|---|--------|------------|--------|
| 0 | AAPL   | 2000-01-03 | NaN    |
| 1 | AAPL   | 2000-01-04 | -0.084 |
| 2 | AAPL   | 2000-01-05 | 0.015  |
| 3 | AAPL   | 2000-01-06 | -0.087 |
| 4 | AAPL   | 2000-01-07 | 0.047  |

The resulting dataframe contains three columns, where the last contains the daily returns (ret). Note that the first entry naturally contains a missing value (NaN) because there is no previous price. Obviously, the use of pct_change() would be meaningless if the time series is not ordered by ascending dates. The function sort_values() provides a convenient way to order observations in the correct way for our application. In case you want to order observations by descending dates, you can use the parameter ascending=False.

For the upcoming examples, we remove missing values, as these would require separate treatment when computing, e.g., sample averages. In general, however, make sure you understand why `NA` values occur and carefully examine if you can simply get rid of these observations. The dataframe `dropna()` method kicks out all rows that contain a missing value in any column.

```
returns = returns.dropna()
```

Next, we visualize the distribution of daily returns in a histogram in Figure 2.2, where we also introduce the `mizani` (Kibirige, 2023a) package for formatting functions. Additionally, we add a dashed line that indicates the five percent quantile of the daily returns to the histogram, which is a (crude) proxy for the worst return of the stock with a probability of at most five percent. The five percent quantile is closely connected to the (historical) value-at-risk, a risk measure commonly monitored by regulators. We refer to Tsay (2010) for a more thorough introduction to stylized facts of returns.

```
from mizani.formatters import percent_format

quantile_05 = returns["ret"].quantile(0.05)

returns_figure = (
  ggplot(returns, aes(x="ret")) +
  geom_histogram(bins=100) +
  geom_vline(aes(xintercept=quantile_05),
               linetype="dashed") +
  labs(x="", y="",
       title="Distribution of daily Apple stock returns") +
  scale_x_continuous(labels=percent_format())
)
returns_figure.draw()
```

Here, `bins=100` determines the number of bins used in the illustration and, hence, implicitly the width of the bins. Before proceeding, make sure you understand how to use the geom `geom_vline()` to add a dashed line that indicates the five percent quantile of the daily returns. A typical task before proceeding with *any* data is to compute summary statistics for the main variables of interest.

```
pd.DataFrame(returns["ret"].describe()).round(3).T
```

|     | count  | mean  | std   | min    | 25%   | 50%   | 75%   | max   |
|-----|--------|-------|-------|--------|-------|-------|-------|-------|
| ret | 5786.0 | 0.001 | 0.025 | -0.519 | -0.01 | 0.001 | 0.013 | 0.139 |

We see that the maximum *daily* return was 13.9 percent. Perhaps not surprisingly, the average daily return is close to but slightly above 0. In line with the illustration above, the large losses on the day with the minimum returns indicate a strong asymmetry in the distribution of returns.

You can also compute these summary statistics for each year individually by imposing `.groupby(returns["date"].dt.year)`, where the call `.dt.year` returns the year of a date variable. More specifically, the few lines of code below compute the summary statistics from above for individual groups of data defined by year. The summary statistics, therefore, allow an eyeball analysis of the time-series dynamics of the return distribution.

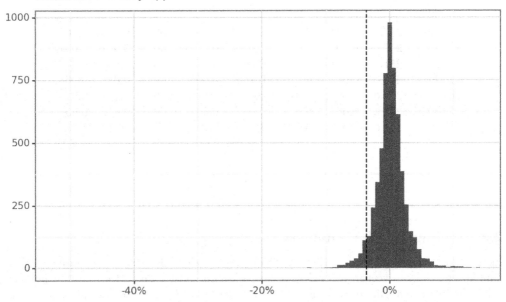

Figure 2.2: The figure shows a histogram of daily Apple stock returns in percent. The dotted vertical line indicates the historical five percent quantile.

```
(returns["ret"]
  .groupby(returns["date"].dt.year)
  .describe()
  .round(3)
)
```

| date | count | mean | std | min | 25% | 50% | 75% | max |
|---|---|---|---|---|---|---|---|---|
| 2000 | 251.0 | -0.003 | 0.055 | -0.519 | -0.034 | -0.002 | 0.027 | 0.137 |
| 2001 | 248.0 | 0.002 | 0.039 | -0.172 | -0.023 | -0.001 | 0.027 | 0.129 |
| 2002 | 252.0 | -0.001 | 0.031 | -0.150 | -0.019 | -0.003 | 0.018 | 0.085 |
| 2003 | 252.0 | 0.002 | 0.023 | -0.081 | -0.012 | 0.002 | 0.015 | 0.113 |
| 2004 | 252.0 | 0.005 | 0.025 | -0.056 | -0.009 | 0.003 | 0.016 | 0.132 |
| 2005 | 252.0 | 0.003 | 0.024 | -0.092 | -0.010 | 0.003 | 0.017 | 0.091 |
| 2006 | 251.0 | 0.001 | 0.024 | -0.063 | -0.014 | -0.002 | 0.014 | 0.118 |
| 2007 | 251.0 | 0.004 | 0.024 | -0.070 | -0.009 | 0.003 | 0.018 | 0.105 |
| 2008 | 253.0 | -0.003 | 0.037 | -0.179 | -0.024 | -0.001 | 0.019 | 0.139 |
| 2009 | 252.0 | 0.004 | 0.021 | -0.050 | -0.009 | 0.002 | 0.015 | 0.068 |
| 2010 | 252.0 | 0.002 | 0.017 | -0.050 | -0.006 | 0.002 | 0.011 | 0.077 |
| 2011 | 252.0 | 0.001 | 0.017 | -0.056 | -0.009 | 0.001 | 0.011 | 0.059 |
| 2012 | 250.0 | 0.001 | 0.019 | -0.064 | -0.008 | 0.000 | 0.012 | 0.089 |
| 2013 | 252.0 | 0.000 | 0.018 | -0.124 | -0.009 | -0.000 | 0.011 | 0.051 |
| 2014 | 252.0 | 0.001 | 0.014 | -0.080 | -0.006 | 0.001 | 0.010 | 0.082 |
| 2015 | 252.0 | 0.000 | 0.017 | -0.061 | -0.009 | -0.001 | 0.009 | 0.057 |

| date | count | mean | std | min | 25% | 50% | 75% | max |
|------|-------|------|-----|-----|-----|-----|-----|-----|
| 2016 | 252.0 | 0.001 | 0.015 | -0.066 | -0.006 | 0.001 | 0.008 | 0.065 |
| 2017 | 251.0 | 0.002 | 0.011 | -0.039 | -0.004 | 0.001 | 0.007 | 0.061 |
| 2018 | 251.0 | -0.000 | 0.018 | -0.066 | -0.009 | 0.001 | 0.009 | 0.070 |
| 2019 | 252.0 | 0.003 | 0.016 | -0.100 | -0.005 | 0.003 | 0.012 | 0.068 |
| 2020 | 253.0 | 0.003 | 0.029 | -0.129 | -0.010 | 0.002 | 0.017 | 0.120 |
| 2021 | 252.0 | 0.001 | 0.016 | -0.042 | -0.008 | 0.001 | 0.012 | 0.054 |
| 2022 | 251.0 | -0.001 | 0.022 | -0.059 | -0.016 | -0.001 | 0.014 | 0.089 |

## 2.2   Scaling Up the Analysis

As a next step, we generalize the code from before such that all the computations can handle an arbitrary vector of stock symbols (e.g., all constituents of an index). Following tidy principles, it is quite easy to download the data, plot the price time series, and tabulate the summary statistics for an arbitrary number of assets.

This is where the magic starts: tidy data makes it extremely easy to generalize the computations from before to as many assets as you like. The following code takes any vector of symbols, e.g., `symbol=["AAPL", "MMM", "BA"]`, and automates the download as well as the plot of the price time series. In the end, we create the table of summary statistics for an arbitrary number of assets. We perform the analysis with data from all current constituents of the Dow Jones Industrial Average index.[1]

We first download a table with DOW Jones constituents from an external website.

```
url = ("https://www.ssga.com/us/en/institutional/etfs/library-content/"
       "products/fund-data/etfs/us/holdings-daily-us-en-dia.xlsx")

symbols = (pd.read_excel(url, skiprows=4, nrows=30)
  .get("Ticker")
  .tolist()
)
```

Next, we can use `yf.download()` to download prices for all stock symbols in the above list and again chain a couple of `pandas` dataframe functions to create a tidy dataset.

```
index_prices = (yf.download(
    tickers=symbols,
    start="2000-01-01",
    end="2022-12-31",
    progress=False
  )
  .melt(ignore_index=False, var_name=["variable", "symbol"])
  .reset_index()
  .pivot(index=["Date", "symbol"], columns="variable", values="value")
```

---

[1] https://en.wikipedia.org/wiki/Dow_Jones_Industrial_Average

```
  .reset_index()
  .rename(columns={
    "Date": "date",
    "Open": "open",
    "High": "high",
    "Low": "low",
    "Close": "close",
    "Adj Close": "adjusted",
    "Volume": "volume"}
  )
)
```

The resulting dataframe contains 173,610 daily observations for 30 different stock symbols. Figure 2.3 illustrates the time series of downloaded *adjusted* prices for each of the constituents of the Dow Jones index. We again draw on the `mizani` package, but this time we use its useful date formatting function to get nicer axis labels. Make sure you understand every single line of code! What are the arguments of `aes()`? Which alternative `geoms` could you use to visualize the time series? Hint: If you do not know the answers, try to change the code to see what difference your intervention causes.

```
from mizani.breaks import date_breaks
from mizani.formatters import date_format

index_prices_figure = (
  ggplot(index_prices,
         aes(y="adjusted", x="date", color="symbol")) +
  geom_line() +
  labs(x="", y="", color="",
      title="Stock prices of DOW index constituents") +
  theme(legend_position="none") +
  scale_x_datetime(date_breaks="5 years", date_labels="%Y")
)
index_prices_figure.draw()
```

Do you notice the small differences relative to the code we used before? `yf.download(symbols)` returns a dataframe for several symbols as well. All we need to do to illustrate all symbols simultaneously is to include `color="symbol"` in the `ggplot` aesthetics. In this way, we generate a separate line for each symbol. Of course, there are simply too many lines on this graph to identify the individual stocks properly, but it illustrates the point well.

The same holds for stock returns. Before computing the returns, we use `groupby("symbol")` such that the `assign()` command is performed to calculate the returns for each symbol individually and assign it to the variable `ret` in the dataframe. The same logic also applies to the computation of summary statistics: `groupby("symbol")` is the key to aggregating the time series into symbol-specific variables of interest.

```
all_returns = (index_prices
  .assign(ret=lambda x: x.groupby("symbol")["adjusted"].pct_change())
  .get(["symbol", "date", "ret"])
  .dropna(subset="ret")
)
```

Stock prices of DOW index constituents

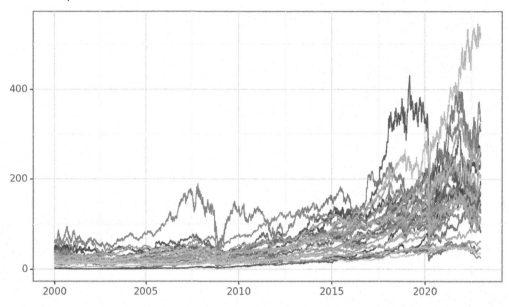

Figure 2.3: The figure shows the stock prices of DOW index constituents. Prices are in USD, adjusted for dividend payments and stock splits.

```
(all_returns
  .groupby("symbol")["ret"]
  .describe()
  .round(3)
)
```

| symbol | count | mean | std | min | 25% | 50% | 75% | max |
|---|---|---|---|---|---|---|---|---|
| AAPL | 5786.0 | 0.001 | 0.025 | -0.519 | -0.010 | 0.001 | 0.013 | 0.139 |
| AMGN | 5786.0 | 0.000 | 0.020 | -0.134 | -0.009 | 0.000 | 0.009 | 0.151 |
| AMZN | 5786.0 | 0.001 | 0.032 | -0.248 | -0.012 | 0.000 | 0.014 | 0.345 |
| AXP | 5786.0 | 0.001 | 0.023 | -0.176 | -0.009 | 0.000 | 0.010 | 0.219 |
| BA | 5786.0 | 0.001 | 0.022 | -0.238 | -0.010 | 0.001 | 0.011 | 0.243 |
| CAT | 5786.0 | 0.001 | 0.020 | -0.145 | -0.010 | 0.001 | 0.011 | 0.147 |
| CRM | 4664.0 | 0.001 | 0.027 | -0.271 | -0.012 | 0.000 | 0.014 | 0.260 |
| CSCO | 5786.0 | 0.000 | 0.024 | -0.162 | -0.009 | 0.000 | 0.010 | 0.244 |
| CVX | 5786.0 | 0.001 | 0.018 | -0.221 | -0.008 | 0.001 | 0.009 | 0.227 |
| DIS | 5786.0 | 0.000 | 0.019 | -0.184 | -0.009 | 0.000 | 0.009 | 0.160 |
| DOW | 954.0 | 0.001 | 0.026 | -0.217 | -0.012 | 0.000 | 0.014 | 0.209 |
| GS | 5786.0 | 0.001 | 0.023 | -0.190 | -0.010 | 0.000 | 0.011 | 0.265 |
| HD | 5786.0 | 0.001 | 0.019 | -0.287 | -0.008 | 0.001 | 0.009 | 0.141 |
| HON | 5786.0 | 0.001 | 0.019 | -0.174 | -0.008 | 0.001 | 0.009 | 0.282 |
| IBM | 5786.0 | 0.000 | 0.017 | -0.155 | -0.007 | 0.000 | 0.008 | 0.120 |
| INTC | 5786.0 | 0.000 | 0.024 | -0.220 | -0.010 | 0.000 | 0.011 | 0.201 |

| | count | mean | std | min | 25% | 50% | 75% | max |
|---|---|---|---|---|---|---|---|---|
| **symbol** | | | | | | | | |
| JNJ | 5786.0 | 0.000 | 0.012 | -0.158 | -0.005 | 0.000 | 0.006 | 0.122 |
| JPM | 5786.0 | 0.001 | 0.024 | -0.207 | -0.009 | 0.000 | 0.010 | 0.251 |
| KO | 5786.0 | 0.000 | 0.013 | -0.101 | -0.005 | 0.000 | 0.006 | 0.139 |
| MCD | 5786.0 | 0.001 | 0.015 | -0.159 | -0.006 | 0.001 | 0.007 | 0.181 |
| MMM | 5786.0 | 0.000 | 0.015 | -0.129 | -0.006 | 0.000 | 0.008 | 0.126 |
| MRK | 5786.0 | 0.000 | 0.017 | -0.268 | -0.007 | 0.000 | 0.008 | 0.130 |
| MSFT | 5786.0 | 0.001 | 0.019 | -0.156 | -0.008 | 0.000 | 0.009 | 0.196 |
| NKE | 5786.0 | 0.001 | 0.019 | -0.198 | -0.008 | 0.001 | 0.009 | 0.155 |
| PG | 5786.0 | 0.000 | 0.013 | -0.302 | -0.005 | 0.000 | 0.006 | 0.120 |
| TRV | 5786.0 | 0.001 | 0.018 | -0.208 | -0.007 | 0.001 | 0.008 | 0.256 |
| UNH | 5786.0 | 0.001 | 0.020 | -0.186 | -0.008 | 0.001 | 0.010 | 0.348 |
| V | 3723.0 | 0.001 | 0.019 | -0.136 | -0.008 | 0.001 | 0.010 | 0.150 |
| VZ | 5786.0 | 0.000 | 0.015 | -0.118 | -0.007 | 0.000 | 0.007 | 0.146 |
| WMT | 5786.0 | 0.000 | 0.015 | -0.114 | -0.007 | 0.000 | 0.007 | 0.117 |

## 2.3 Other Forms of Data Aggregation

Of course, aggregation across variables other than `symbol` can also make sense. For instance, suppose you are interested in answering the question: Are days with high aggregate trading volume likely followed by days with high aggregate trading volume? To provide some initial analysis on this question, we take the downloaded data and compute aggregate daily trading volume for all Dow Jones constituents in USD. Recall that the column `volume` is denoted in the number of traded shares. Thus, we multiply the trading volume with the daily adjusted closing price to get a proxy for the aggregate trading volume in USD. Scaling by `1e9` (Python can handle scientific notation) denotes daily trading volume in billion USD.

```
trading_volume = (index_prices
    .assign(trading_volume=lambda x: (x["volume"]*x["adjusted"])/1e9)
    .groupby("date")["trading_volume"]
    .sum()
    .reset_index()
    .assign(trading_volume_lag=lambda x: x["trading_volume"].shift(periods=1))
)

trading_volume_figure = (
  ggplot(trading_volume,
         aes(x="date", y="trading_volume")) +
  geom_line() +
  labs(x="", y="",
       title=("Aggregate daily trading volume of DOW index constituents "
              "in billion USD")) +
  scale_x_datetime(date_breaks="5 years", date_labels="%Y")
)
trading_volume_figure.draw()
```

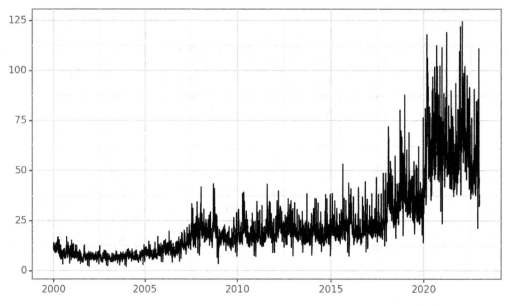

Figure 2.4: The figure shows the total daily trading volume in billion USD.

Figure 2.4 indicates a clear upward trend in aggregated daily trading volume. In particular, since the outbreak of the COVID-19 pandemic, markets have processed substantial trading volumes, as analyzed, for instance, by Goldstein et al. (2021). One way to illustrate the persistence of trading volume would be to plot volume on day $t$ against volume on day $t-1$ as in the example below. In Figure 2.5, we add a dotted 45°-line to indicate a hypothetical one-to-one relation by `geom_abline()`, addressing potential differences in the axes' scales.

```
trading_volume_figure = (
  ggplot(trading_volume,
         aes(x="trading_volume_lag", y="trading_volume")) +
  geom_point() +
  geom_abline(aes(intercept=0, slope=1), linetype="dashed") +
  labs(x="Previous day aggregate trading volume",
       y="Aggregate trading volume",
       title=("Persistence in daily trading volume of DOW constituents "
              "in billion USD"))
)
trading_volume_figure.draw()
```

## 2.4   Portfolio Choice Problems

In the previous part, we show how to download stock market data and inspect it with graphs and summary statistics. Now, we move to a typical question in Finance: How to allocate wealth across different assets optimally. The standard framework for optimal portfolio

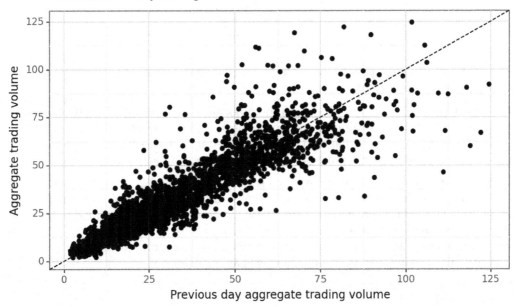

Figure 2.5: The figure a scatterplot of aggregate trading volume against previous-day aggregate trading volume.

selection considers investors that prefer higher future returns but dislike future return volatility (defined as the square root of the return variance, i.e., the risk): the *mean-variance investor* (Markowitz, 1952).

An essential tool to evaluate portfolios in the mean-variance context is the *efficient frontier*, the set of portfolios that satisfies the condition that no other portfolio exists with a higher expected return but with the same volatility, see, e.g., Merton (1972). We compute and visualize the efficient frontier for several stocks. First, we extract each asset's *monthly* returns. In order to keep things simple, we work with a balanced panel and exclude DOW constituents for which we do not observe a price on every single trading day since the year 2000.

```
prices = (index_prices
  .groupby("symbol")
  .apply(lambda x: x.assign(counts=x["adjusted"].dropna().count()))
  .reset_index(drop=True)
  .query("counts == counts.max()")
)
```

Next, we transform the returns from a tidy dataframe into a $(T \times N)$ matrix with one column for each of the $N$ symbols and one row for each of the $T$ trading days to compute the sample average return vector

$$\hat{\mu} = \frac{1}{T} \sum_{t=1}^{T} r_t, \tag{2.2}$$

where $r_t$ is the $N$ vector of returns on date $t$ and the sample covariance matrix

$$\hat{\Sigma} = \frac{1}{T-1} \sum_{t=1}^{T} (r_t - \hat{\mu})(r_t - \hat{\mu})'. \tag{2.3}$$

We achieve this by using `pivot()` with the new column names from the column `symbol` and setting the values to `adjusted`.

In financial econometrics, a core focus falls on problems that arise if the investor has to rely on estimates $\hat{\mu}$ and $\hat{\Sigma}$ instead of using the vector of expected returns $\mu$ and the variance-covariance matrix $\Sigma$. We highlight the impact of estimation uncertainty on the portfolio performance in various backtesting applications in Chapter 17 and Chapter 18.

For now, we focus on a much more restricted set of assumptions: The $N$ assets are fixed, and the first two moments of the distribution of the returns are determined by the parameters $\mu$ and $\Sigma$. Thus, even though we proceed with the vector of sample average returns and the sample variance-covariance matrix, those will be handled as the *true* parameters of the return distribution for the rest of this chapter. We, therefore, refer to $\Sigma$ and $\mu$ instead of explicitly highlighting that the sample moments are estimates.

```
returns_matrix = (prices
  .pivot(columns="symbol", values="adjusted", index="date")
  .resample("m")
  .last()
  .pct_change()
  .dropna()
)
mu = np.array(returns_matrix.mean()).T
sigma = np.array(returns_matrix.cov())
```

Then, we compute the minimum variance portfolio weights $\omega_{\mathrm{mvp}}$ as well as the expected return $\omega_{\mathrm{mvp}}'\mu$ and volatility $\sqrt{\omega_{\mathrm{mvp}}'\Sigma\omega_{\mathrm{mvp}}}$ of this portfolio. Recall that the minimum variance portfolio is the vector of portfolio weights that are the solution to

$$\omega_{\mathrm{mvp}} = \arg\min \omega'\Sigma\omega \text{ s.t. } \sum_{i=1}^{N} \omega_i = 1. \tag{2.4}$$

The constraint that weights sum up to one simply implies that all funds are distributed across the available asset universe, i.e., there is no possibility to retain cash. It is easy to show analytically that $\omega_{\mathrm{mvp}} = \frac{\Sigma^{-1}\iota}{\iota'\Sigma^{-1}\iota}$, where $\iota$ is a vector of ones and $\Sigma^{-1}$ is the inverse of $\Sigma$. We provide the proof of the analytical solution in Appendix B.

```
N = returns_matrix.shape[1]
iota = np.ones(N)
sigma_inv = np.linalg.inv(sigma)

mvp_weights = sigma_inv @ iota
mvp_weights = mvp_weights/mvp_weights.sum()
mvp_return = mu.T @ mvp_weights
mvp_volatility = np.sqrt(mvp_weights.T @ sigma @ mvp_weights)
mvp_moments = pd.DataFrame({"value": [mvp_return, mvp_volatility]},
                           index=["average_ret", "volatility"])
mvp_moments.round(3)
```

|  | value |
|---|---|
| average_ret | 0.008 |
| volatility | 0.032 |

The command `np.linalg.inv()` returns the inverse of a matrix such that `np.linalg.inv(sigma)` delivers $\Sigma^{-1}$ (if a unique solution exists).

Note that the *monthly* volatility of the minimum variance portfolio is of the same order of magnitude as the *daily* standard deviation of the individual components. Thus, the diversification benefits in terms of risk reduction are tremendous!

Next, we set out to find the weights for a portfolio that achieves, as an example, three times the expected return of the minimum variance portfolio. However, mean-variance investors are not interested in any portfolio that achieves the required return but rather in the efficient portfolio, i.e., the portfolio with the lowest standard deviation. If you wonder where the solution $\omega_{\text{eff}}$ comes from: The efficient portfolio is chosen by an investor who aims to achieve minimum variance *given a minimum acceptable expected return* $\bar{\mu}$. Hence, their objective function is to choose $\omega_{\text{eff}}$ as the solution to

$$\omega_{\text{eff}}(\bar{\mu}) = \arg\min \omega'\Sigma\omega \text{ s.t. } \omega'\iota = 1 \text{ and } \omega'\mu \geq \bar{\mu}. \tag{2.5}$$

In Appendix B, we show that the efficient portfolio takes the form (for $\bar{\mu} \geq D/C = \mu'\omega_{\text{mvp}}$)

$$\omega_{\text{eff}}(\bar{\mu}) = \omega_{\text{mvp}} + \frac{\tilde{\lambda}}{2}\left(\Sigma^{-1}\mu - \frac{D}{C}\Sigma^{-1}\iota\right)$$

where $C := \iota'\Sigma^{-1}\iota$, $D := \iota'\Sigma^{-1}\mu$, $E := \mu'\Sigma^{-1}\mu$, and $\tilde{\lambda} = 2\frac{\bar{\mu}-D/C}{E-D^2/C}$.

The code below implements the analytic solution to this optimization problem for a benchmark return $\bar{\mu}$, which we set to 3 times the expected return of the minimum variance portfolio. We encourage you to verify that it is correct.

```
benchmark_multiple = 3
mu_bar = benchmark_multiple*mvp_return
C = iota.T @ sigma_inv @ iota
D = iota.T @ sigma_inv @ mu
E = mu.T @ sigma_inv @ mu
lambda_tilde = 2*(mu_bar-D/C)/(E-D**2/C)
efp_weights = mvp_weights+lambda_tilde/2*(sigma_inv @ mu-D*mvp_weights)
```

## 2.5 The Efficient Frontier

The mutual fund separation theorem states that as soon as we have two efficient portfolios (such as the minimum variance portfolio $\omega_{\text{mvp}}$ and the efficient portfolio for a higher required level of expected returns $\omega_{\text{eff}}(\bar{\mu})$), we can characterize the entire efficient frontier by combining these two portfolios. That is, any linear combination of the two portfolio weights will again represent an efficient portfolio. The code below implements the construction of the *efficient frontier*, which characterizes the highest expected return achievable at each level of risk. To understand the code better, make sure to familiarize yourself with the inner workings of the `for` loop.

```
length_year = 12
a = np.arange(-0.4, 2.0, 0.01)
res = pd.DataFrame(columns=["mu", "sd"], index=a).astype(float)

for i in a:
    w = (1-i)*mvp_weights+i*efp_weights
    res.loc[i, "mu"] = (w.T @ mu)*length_year
    res.loc[i, "sd"] = np.sqrt(w.T @ sigma @ w)*np.sqrt(length_year)
```

The code above proceeds in two steps: First, we compute a vector of combination weights $a$, and then we evaluate the resulting linear combination with $a \in \mathbb{R}$:

$$\omega^* = a\omega_{\text{eff}}(\bar{\mu}) + (1-a)\omega_{\text{mvp}} = \omega_{\text{mvp}} + \frac{\lambda^*}{2}\left(\Sigma^{-1}\mu - \frac{D}{C}\Sigma^{-1}\iota\right) \qquad (2.6)$$

with $\lambda^* = 2\frac{a\bar{\mu}+(1-a)\tilde{\mu}-D/C}{E-D^2/C}$. It follows that $\omega^* = \omega_{\text{eff}}(a\bar{\mu} + (1-a)\tilde{\mu})$, in other words, $\omega^*$ is an efficient portfolio that proofs the mutual fund separation theorem.

Finally, it is simple to visualize the efficient frontier alongside the two efficient portfolios within one powerful figure using the `ggplot` function from **plotnine** (see Figure 2.6). We also add the individual stocks in the same call. We compute annualized returns based on the simple assumption that monthly returns are independent and identically distributed. Thus, the average annualized return is just twelve times the expected monthly return.

```
mvp_return = (mu.T @ mvp_weights)*length_year
mvp_volatility = (np.sqrt(mvp_weights.T @ sigma @ mvp_weights)*
                np.sqrt(length_year))
efp_return = mu_bar*length_year
efp_volatility = (np.sqrt(efp_weights.T @ sigma @ efp_weights)*
                np.sqrt(length_year))

res_figure = (
  ggplot(res, aes(x="sd", y="mu")) +
  geom_point() +
  geom_point(
    pd.DataFrame({"mu": [mvp_return, efp_return],
                "sd": [mvp_volatility, efp_volatility]}),
    size=4
  ) +
  geom_point(
    pd.DataFrame({"mu": mu*length_year,
                "sd": np.sqrt(np.diag(sigma))*np.sqrt(length_year)})
  ) +
  labs(x="Annualized standard deviation",
      y="Annualized expected return",
      title="Efficient frontier for DOW index constituents") +
  scale_x_continuous(labels=percent_format()) +
  scale_y_continuous(labels=percent_format())
)
res_figure.draw()
```

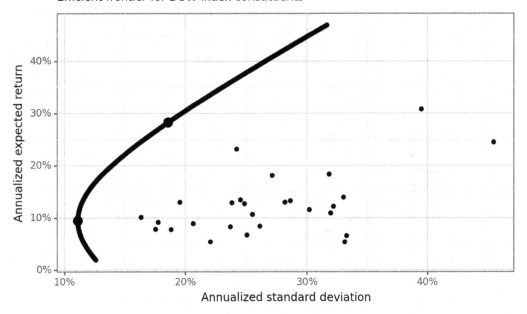

Figure 2.6: The figure shows the efficient frontier for DOW index constituents. The big dots indicate the location of the minimum variance and the efficient portfolio that delivers three times the expected return of the minimum variance portfolio, respectively. The small dots indicate the location of the individual constituents.

The line in Figure 2.6 indicates the efficient frontier: the set of portfolios a mean-variance efficient investor would choose from. Compare the performance relative to the individual assets (the dots); it should become clear that diversifying yields massive performance gains (at least as long as we take the parameters $\Sigma$ and $\mu$ as given).

## 2.6   Exercises

1. Download daily prices for another stock market symbol of your choice from Ya-hoo!Finance with `yf.download()` from the `yfinance` package. Plot two time series of the ticker's unadjusted and adjusted closing prices. Explain the differences.

2. Compute daily net returns for an asset of your choice and visualize the distribution of daily returns in a histogram using 100 bins. Also, use `geom_vline()` to add a dashed red vertical line that indicates the five percent quantile of the daily returns. Compute summary statistics (mean, standard deviation, minimum and maximum) for the daily returns.

3. Take your code from before and generalize it such that you can perform all the computations for an arbitrary vector of tickers (e.g., `ticker = ["AAPL", "MMM", "BA"]`). Automate the download, the plot of the price time series, and create a table of return summary statistics for this arbitrary number of assets.

4. Are days with high aggregate trading volume often also days with large absolute returns? Find an appropriate visualization to analyze the question using the ticker AAPL.

5. Compute monthly returns from the downloaded stock market prices. Compute the vector of historical average returns and the sample variance-covariance matrix. Compute the minimum variance portfolio weights and the portfolio volatility and average returns. Visualize the mean-variance efficient frontier. Choose one of your assets and identify the portfolio which yields the same historical volatility but achieves the highest possible average return.

6. In the portfolio choice analysis, we restricted our sample to all assets trading every day since 2000. How is such a decision a problem when you want to infer future expected portfolio performance from the results?

7. The efficient frontier characterizes the portfolios with the highest expected return for different levels of risk. Identify the portfolio with the highest expected return per standard deviation. Which famous performance measure is close to the ratio of average returns to the standard deviation of returns?

# Part II

# Financial Data

# 3

# Accessing and Managing Financial Data

In this chapter, we suggest a way to organize your financial data. Everybody who has experience with data is also familiar with storing data in various formats like CSV, XLS, XLSX, or other delimited value storage. Reading and saving data can become very cumbersome when using different data formats and across different projects. Moreover, storing data in delimited files often leads to problems with respect to column type consistency. For instance, date-type columns frequently lead to inconsistencies across different data formats and programming languages.

This chapter shows how to import different open-source datasets. Specifically, our data comes from the application programming interface (API) of Yahoo!Finance, a downloaded standard CSV file, an XLSX file stored in a public Google Drive repository, and other macroeconomic time series. We store all the data in a *single* database, which serves as the only source of data in subsequent chapters. We conclude the chapter by providing some tips on managing databases.

First, we load the Python packages that we use throughout this chapter. Later on, we load more packages in the sections where we need them.

```
import pandas as pd
import numpy as np
```

Moreover, we initially define the date range for which we fetch and store the financial data, making future data updates tractable. In case you need another time frame, you can adjust the dates below. Our data starts with 1960 since most asset pricing studies use data from 1962 on.

```
start_date = "1960-01-01"
end_date = "2022-12-31"
```

## 3.1 Fama-French Data

We start by downloading some famous Fama-French factors (e.g., Fama and French, 1993) and portfolio returns commonly used in empirical asset pricing. Fortunately, the `pandas-datareader` package provides a simple interface to read data from Kenneth French's Data Library.

```
import pandas_datareader as pdr
```

We can use the `pdr.DataReader()` function of the package to download monthly Fama-French factors. The set *Fama/French 3 Factors* contains the return time series of the market (`mkt_excess`), size (`smb`), and value (`hml`) factors alongside the risk-free rates (`rf`). Note that we have to do some manual work to parse all the columns correctly and scale them appropriately, as the raw Fama-French data comes in a unique data format. For precise descriptions of the variables, we suggest consulting Prof. Kenneth French's finance data library directly. If you are on the website, check the raw data files to appreciate the time you can save thanks to `pandas_datareader`.

```python
factors_ff3_monthly_raw = pdr.DataReader(
  name="F-F_Research_Data_Factors",
  data_source="famafrench",
  start=start_date,
  end=end_date)[0]

factors_ff3_monthly = (factors_ff3_monthly_raw
  .divide(100)
  .reset_index(names="month")
  .assign(month=lambda x: pd.to_datetime(x["month"].astype(str)))
  .rename(str.lower, axis="columns")
  .rename(columns={"mkt-rf": "mkt_excess"})
)
```

We also download the set *5 Factors (2x3)*, which additionally includes the return time series of the profitability (`rmw`) and investment (`cma`) factors. We demonstrate how the monthly factors are constructed in Chapter 11.

```python
factors_ff5_monthly_raw = pdr.DataReader(
  name="F-F_Research_Data_5_Factors_2x3",
  data_source="famafrench",
  start=start_date,
  end=end_date)[0]

factors_ff5_monthly = (factors_ff5_monthly_raw
  .divide(100)
  .reset_index(names="month")
  .assign(month=lambda x: pd.to_datetime(x["month"].astype(str)))
  .rename(str.lower, axis="columns")
  .rename(columns={"mkt-rf": "mkt_excess"})
)
```

It is straightforward to download the corresponding *daily* Fama-French factors with the same function.

```python
factors_ff3_daily_raw = pdr.DataReader(
  name="F-F_Research_Data_Factors_daily",
  data_source="famafrench",
  start=start_date,
  end=end_date)[0]

factors_ff3_daily = (factors_ff3_daily_raw
  .divide(100)
  .reset_index(names="date")
  .rename(str.lower, axis="columns")
  .rename(columns={"mkt-rf": "mkt_excess"})
)
```

In a subsequent chapter, we also use the monthly returns from ten industry portfolios, so let us fetch that data, too.

```
industries_ff_monthly_raw = pdr.DataReader(
  name="10_Industry_Portfolios",
  data_source="famafrench",
  start=start_date,
  end=end_date)[0]

industries_ff_monthly = (industries_ff_monthly_raw
  .divide(100)
  .reset_index(names="month")
  .assign(month=lambda x: pd.to_datetime(x["month"].astype(str)))
  .rename(str.lower, axis="columns")
)
```

It is worth taking a look at all available portfolio return time series from Kenneth French's homepage. You should check out the other sets by calling `pdr.famafrench.get_available_datasets()`.

---

## 3.2 q-Factors

In recent years, the academic discourse experienced the rise of alternative factor models, e.g., in the form of the Hou et al. (2014) *q*-factor model. We refer to the extended background[1] information provided by the original authors for further information. The *q*-factors can be downloaded directly from the authors' homepage from within `pd.read_csv()`.

We also need to adjust this data. First, we discard information we will not use in the remainder of the book. Then, we rename the columns with the "R_"-prescript using regular expressions and write all column names in lowercase. We then query the data to select observations between the start and end dates. Finally, we use the double asterisk (**) notation in the `assign` function to apply the same transform of dividing by 100 to all four factors by iterating through them. You should always try sticking to a consistent style for naming objects, which we try to illustrate here - the emphasis is on *try*. You can check out style guides available online, e.g., Hadley Wickham's `tidyverse` style guide.[2]

```
factors_q_monthly_link = (
  "https://global-q.org/uploads/1/2/2/6/122679606/"
  "q5_factors_monthly_2022.csv"
)

factors_q_monthly = (pd.read_csv(factors_q_monthly_link)
  .assign(
    month=lambda x: (
      pd.to_datetime(x["year"].astype(str) + "-" +
        x["month"].astype(str) + "-01"))
  )
```

---

[1] http://global-q.org/background.html
[2] https://style.tidyverse.org/index.html

```
 .drop(columns=["R_F", "R_MKT", "year"])
 .rename(columns=lambda x: x.replace("R_", "").lower())
 .query(f"month >= '{start_date}' and month <= '{end_date}'")
 .assign(
   **{col: lambda x: x[col]/100 for col in ["me", "ia", "roe", "eg"]}
 )
)
```

## 3.3   Macroeconomic Predictors

Our next data source is a set of macroeconomic variables often used as predictors for the equity premium. Welch and Goyal (2008) comprehensively reexamine the performance of variables suggested by the academic literature to be good predictors of the equity premium. The authors host the data until 2022 on Amit Goyal's website.[3] Since the data is an XLSX-file stored on a public Google Drive location, we need additional packages to access the data directly from our Python session. Usually, you need to authenticate if you interact with Google drive directly in Python. Since the data is stored via a public link, we can proceed without any authentication.

```
sheet_id = "1g4LOaRj4TvwJr9RIaA_nwrXXWTOy46bP"
sheet_name = "macro_predictors.xlsx"
macro_predictors_link = (
  f"https://docs.google.com/spreadsheets/d/{sheet_id}"
  f"/gviz/tq?tqx=out:csv&sheet={sheet_name}"
)
```

Next, we read in the new data and transform the columns into the variables that we later use:

1. The dividend price ratio (dp), the difference between the log of dividends and the log of prices, where dividends are 12-month moving sums of dividends paid on the S&P 500 index, and prices are monthly averages of daily closing prices (Campbell and Shiller, 1988; Campbell and Yogo, 2006).
2. Dividend yield (dy), the difference between the log of dividends and the log of lagged prices (Ball, 1978).
3. Earnings price ratio (ep), the difference between the log of earnings and the log of prices, where earnings are 12-month moving sums of earnings on the S&P 500 index (Campbell and Shiller, 1988).
4. Dividend payout ratio (de), the difference between the log of dividends and the log of earnings (Lamont, 1998).
5. Stock variance (svar), the sum of squared daily returns on the S&P 500 index (Guo, 2006).
6. Book-to-market ratio (bm), the ratio of book value to market value for the Dow Jones Industrial Average (Kothari and Shanken, 1997).

---

[3]https://sites.google.com/view/agoyal145

7. Net equity expansion (`ntis`), the ratio of 12-month moving sums of net issues by NYSE listed stocks divided by the total end-of-year market capitalization of NYSE stocks (Campbell et al., 2008).

8. Treasury bills (`tbl`), the 3-Month Treasury Bill: Secondary Market Rate from the economic research database at the Federal Reserve Bank at St. Louis (Campbell, 1987).

9. Long-term yield (`lty`), the long-term government bond yield from Ibbotson's Stocks, Bonds, Bills, and Inflation Yearbook (Welch and Goyal, 2008).

10. Long-term rate of returns (`ltr`), the long-term government bond returns from Ibbotson's Stocks, Bonds, Bills, and Inflation Yearbook (Welch and Goyal, 2008).

11. Term spread (`tms`), the difference between the long-term yield on government bonds and the Treasury bill (Campbell, 1987).

12. Default yield spread (`dfy`), the difference between BAA and AAA-rated corporate bond yields (Fama and French, 1989).

13. Inflation (`infl`), the Consumer Price Index (All Urban Consumers) from the Bureau of Labor Statistics (Campbell and Vuolteenaho, 2004).

For variable definitions and the required data transformations, you can consult the material on Amit Goyal's website.[4]

```
macro_predictors = (
  pd.read_csv(macro_predictors_link, thousands=",")
  .assign(
    month=lambda x: pd.to_datetime(x["yyyymm"], format="%Y%m"),
    dp=lambda x: np.log(x["D12"])-np.log(x["Index"]),
    dy=lambda x: np.log(x["D12"])-np.log(x["D12"].shift(1)),
    ep=lambda x: np.log(x["E12"])-np.log(x["Index"]),
    de=lambda x: np.log(x["D12"])-np.log(x["E12"]),
    tms=lambda x: x["lty"]-x["tbl"],
    dfy=lambda x: x["BAA"]-x["AAA"]
  )
  .rename(columns={"b/m": "bm"})
  .get(["month", "dp", "dy", "ep", "de", "svar", "bm",
        "ntis", "tbl", "lty", "ltr", "tms", "dfy", "infl"])
  .query("month >= @start_date and month <= @end_date")
  .dropna()
)
```

## 3.4 Other Macroeconomic Data

The Federal Reserve bank of St. Louis provides the Federal Reserve Economic Data (FRED), an extensive database for macroeconomic data. In total, there are 817,000 US and international time series from 108 different sources. As an illustration, we use the already familiar `pandas-datareader` package to fetch consumer price index (CPI) data that can be found under the CPIAUCNS[5] key.

---

[4]https://sites.google.com/view/agoyal145
[5]https://fred.stlouisfed.org/series/CPIAUCNS

```
cpi_monthly = (pdr.DataReader(
    name="CPIAUCNS",
    data_source="fred",
    start=start_date,
    end=end_date
  )
  .reset_index(names="month")
  .rename(columns={"CPIAUCNS": "cpi"})
  .assign(cpi=lambda x: x["cpi"]/x["cpi"].iloc[-1])
)
```

Note that we use the `assign()` in the last line to set the current (latest) price level as the reference inflation level. To download other time series, we just have to look it up on the FRED website and extract the corresponding key from the address. For instance, the producer price index for gold ores can be found under the PCU21222121222210[6] key.

## 3.5   Setting Up a Database

Now that we have downloaded some (freely available) data from the web into the memory of our Python session, let us set up a database to store that information for future use. We will use the data stored in this database throughout the following chapters, but you could alternatively implement a different strategy and replace the respective code.

There are many ways to set up and organize a database, depending on the use case. For our purpose, the most efficient way is to use an SQLite[7]-database, which is the C-language library that implements a small, fast, self-contained, high-reliability, full-featured SQL database engine. Note that SQL[8] (Structured Query Language) is a standard language for accessing and manipulating databases.

```
import sqlite3
```

An SQLite-database is easily created - the code below is really all there is. You do not need any external software. Otherwise, date columns are stored and retrieved as integers. We will use the resulting file `tidy_finance.db` in the subfolder `data` for all subsequent chapters to retrieve our data.

```
tidy_finance = sqlite3.connect(database="data/tidy_finance_python.sqlite")
```

Next, we create a remote table with the monthly Fama-French factor data. We do so with the `pandas` function `to_sql()`, which copies the data to our SQLite-database.

```
(factors_ff3_monthly
  .to_sql(name="factors_ff3_monthly",
          con=tidy_finance,
          if_exists="replace",
          index=False)
)
```

---

[6]https://fred.stlouisfed.org/series/PCU21222121222210
[7]https://SQLite.org/
[8]https://en.wikipedia.org/wiki/SQL

Now, if we want to have the whole table in memory, we need to call `pd.read_sql_query()` with the corresponding query. You will see that we regularly load the data into the memory in the next chapters.

```
pd.read_sql_query(
  sql="SELECT month, rf FROM factors_ff3_monthly",
  con=tidy_finance,
  parse_dates={"month"}
)
```

|     | month      | rf     |
| --- | ---------- | ------ |
| 0   | 1960-01-01 | 0.0033 |
| 1   | 1960-02-01 | 0.0029 |
| 2   | 1960-03-01 | 0.0035 |
| 3   | 1960-04-01 | 0.0019 |
| 4   | 1960-05-01 | 0.0027 |
| ... | ...        | ...    |
| 751 | 2022-08-01 | 0.0019 |
| 752 | 2022-09-01 | 0.0019 |
| 753 | 2022-10-01 | 0.0023 |
| 754 | 2022-11-01 | 0.0029 |
| 755 | 2022-12-01 | 0.0033 |

The last couple of code chunks are really all there is to organizing a simple database! You can also share the SQLite database across devices and programming languages.

Before we move on to the next data source, let us also store the other six tables in our new SQLite database.

```
data_dict = {
  "factors_ff5_monthly": factors_ff5_monthly,
  "factors_ff3_daily": factors_ff3_daily,
  "industries_ff_monthly": industries_ff_monthly,
  "factors_q_monthly": factors_q_monthly,
  "macro_predictors": macro_predictors,
  "cpi_monthly": cpi_monthly
}

for key, value in data_dict.items():
    value.to_sql(name=key,
                 con=tidy_finance,
                 if_exists="replace",
                 index=False)
```

From now on, all you need to do to access data that is stored in the database is to follow two steps: (i) Establish the connection to the SQLite-database and (ii) execute the query to fetch the data. For your convenience, the following steps show all you need in a compact fashion.

```
import pandas as pd
import sqlite3
```

```
tidy_finance = sqlite3.connect(database="data/tidy_finance_python.sqlite")

factors_q_monthly = pd.read_sql_query(
  sql="SELECT * FROM factors_q_monthly",
  con=tidy_finance,
  parse_dates={"month"}
)
```

## 3.6    Managing SQLite Databases

Finally, at the end of our data chapter, we revisit the SQLite database itself. When you drop database objects such as tables or delete data from tables, the database file size remains unchanged because SQLite just marks the deleted objects as free and reserves their space for future uses. As a result, the database file always grows in size.

To optimize the database file, you can run the `VACUUM` command in the database, which rebuilds the database and frees up unused space. You can execute the command in the database using the `execute()` function.

```
tidy_finance.execute("VACUUM")
```

The `VACUUM` command actually performs a couple of additional cleaning steps, which you can read about in this tutorial.[9]

## 3.7    Exercises

1. Download the monthly Fama-French factors manually from Kenneth French's data library[10] and read them in via `pd.read_csv()`. Validate that you get the same data as via the **pandas-datareader** package.
2. Download the daily Fama-French 5 factors using the `pdr.DataReader()` package. After the successful download and conversion to the column format that we used above, compare the `rf`, `mkt_excess`, `smb`, and `hml` columns of `factors_ff3_daily` to `factors_ff5_daily`. Discuss any differences you might find.

---

[9]https://www.sqlitetutorial.net/sqlite-vacuum/
[10]https://mba.tuck.dartmouth.edu/pages/faculty/ken.french/data_library.html

# 4

# *WRDS, CRSP, and Compustat*

This chapter shows how to connect to Wharton Research Data Services (WRDS),[1] a popular provider of financial and economic data for research applications. We use this connection to download the most commonly used data for stock and firm characteristics, CRSP and Compustat. Unfortunately, this data is not freely available, but most students and researchers typically have access to WRDS through their university libraries. Assuming that you have access to WRDS, we show you how to prepare and merge the databases and store them in the SQLite database introduced in the previous chapter. We conclude this chapter by providing some tips for working with the WRDS database.

If you don't have access to WRDS but still want to run the code in this book, we refer to Appendix C, where we show how to create a dummy database that contains the WRDS tables and corresponding columns. With this database at hand, all code chunks in this book can be executed with this dummy database.

First, we load the Python packages that we use throughout this chapter. Later on, we load more packages in the sections where we need them. The last two packages are used for plotting.

```
import pandas as pd
import numpy as np
import sqlite3

from plotnine import *
from mizani.formatters import comma_format, percent_format
from datetime import datetime
```

We use the same date range as in the previous chapter to ensure consistency. However, we have to use the date format that the WRDS database expects.

```
start_date = "01/01/1960"
end_date = "12/31/2022"
```

## 4.1 Accessing WRDS

WRDS is the most widely used source for asset and firm-specific financial data used in academic settings. WRDS is a data platform that provides data validation, flexible delivery options, and access to many different data sources. The data at WRDS is also organized

---

[1] https://wrds-www.wharton.upenn.edu/

in an SQL database, although they use the PostgreSQL[2] engine. This database engine is just as easy to handle with Python as SQL. We use the `sqlalchemy` package to establish a connection to the WRDS database because it already contains a suitable driver.[3]

```
from sqlalchemy import create_engine
```

To establish a connection, you use the function `create_engine()` with a connection string. Note that you need to replace the `WRDS_USER` and `WRDS_PASSWORD` arguments with your own credentials. We defined environment variables for the purpose of this book because we obviously do not want (and are not allowed) to share our credentials with the rest of the world (these environment variables are stored in the `.env` file of our project directory, loaded with `load_dotenv()`, and then called with the `os.getenv()` function).

Additionally, you have to use two-factor authentication since May 2023 when establishing a remote connection to WRDS. You have two choices to provide the additional identification. First, if you have Duo Push enabled for your WRDS account, you will receive a push notification on your mobile phone when trying to establish a connection with the code below. Upon accepting the notification, you can continue your work. Second, you can log in to a WRDS website that requires two-factor authentication with your username and the same IP address. Once you have successfully identified yourself on the website, your username-IP combination will be remembered for 30 days, and you can comfortably use the remote connection below.

```
import os
from dotenv import load_dotenv
load_dotenv()

connection_string = (
  "postgresql+psycopg2://"
 f"{os.getenv('WRDS_USER')}:{os.getenv('WRDS_PASSWORD')}"
  "@wrds-pgdata.wharton.upenn.edu:9737/wrds"
)

wrds = create_engine(connection_string, pool_pre_ping=True)
```

The remote connection to WRDS is very useful. Yet, the database itself contains many different tables. You can check the WRDS homepage[4] to identify the table's name you are looking for (if you go beyond our exposition).

## 4.2   Downloading and Preparing CRSP

The Center for Research in Security Prices (CRSP)[5] provides the most widely used data for US stocks. We use the `wrds` engine object that we just created to first access monthly CRSP return data. Actually, we need two tables to get the desired data: (i) the CRSP monthly security file (`msf`), and (ii) the historical identifying information (`stksecurityinfohist`).

---

[2]https://www.postgresql.org/

[3]An alternative to establish a connection to WRDS is to use the `WRDS-Py` library. We chose to work with `sqlalchemy` (Bayer, 2012) to show how to access PostgreSQL engines in general.

[4]https://wrds-www.wharton.upenn.edu/

[5]https://crsp.org/

We use the two remote tables to fetch the data we want to put into our local database. Just as above, the idea is that we let the WRDS database do all the work and just download the data that we actually need. We apply common filters and data selection criteria to narrow down our data of interest: (i) we keep only data in the time windows of interest, (ii) we keep only US-listed stocks as identified via no special share types (`sharetype = 'NS'`), security type equity (`securitytype = 'EQTY'`), security sub type common stock (`securitysubtype = 'COM'`), issuers that are a corporation (`issuertype %in% c("ACOR", "CORP")`), and (iii) we keep only months within permno-specific start dates (`secinfostartdt`) and end dates (`secinfoenddt`). As of July 2022, there is no need to additionally download delisting information since it is already contained in the most recent version of `msf`.

```
crsp_monthly_query = (
  "SELECT msf.permno, msf.mthcaldt AS date, "
      "date_trunc('month', msf.mthcaldt)::date AS month, "
      "msf.mthret AS ret, msf.shrout, msf.mthprc AS altprc, "
      "msf.primaryexch, msf.siccd "
    "FROM crsp.msf_v2 AS msf "
    "LEFT JOIN crsp.stksecurityinfohist AS ssih "
    "ON msf.permno = ssih.permno AND "
      "ssih.secinfostartdt <= msf.mthcaldt AND "
      "msf.mthcaldt <= ssih.secinfoenddt "
  f"WHERE msf.mthcaldt BETWEEN '{start_date}' AND '{end_date}' "
      "AND ssih.sharetype = 'NS' "
      "AND ssih.securitytype = 'EQTY' "
      "AND ssih.securitysubtype = 'COM' "
      "AND ssih.usincflg = 'Y' "
      "AND ssih.issuertype in ('ACOR', 'CORP')"
)

crsp_monthly = (pd.read_sql_query(
    sql=crsp_monthly_query,
    con=wrds,
    dtype={"permno": int, "siccd": int},
    parse_dates={"date", "month"})
  .assign(shrout=lambda x: x["shrout"]*1000)
)
```

Now, we have all the relevant monthly return data in memory and proceed with preparing the data for future analyses. We perform the preparation step at the current stage since we want to avoid executing the same mutations every time we use the data in subsequent chapters.

The first additional variable we create is market capitalization (`mktcap`), which is the product of the number of outstanding shares (`shrout`) and the last traded price in a month (`altprc`). Note that in contrast to returns (`ret`), these two variables are not adjusted ex-post for any corporate actions like stock splits. We also keep the market cap in millions of USD just for convenience, as we do not want to print huge numbers in our figures and tables. In addition, we set zero market capitalization to missing as it makes conceptually little sense (i.e., the firm would be bankrupt).

```
crsp_monthly = (crsp_monthly
  .assign(mktcap=lambda x: x["shrout"]*x["altprc"]/1000000)
```

```
  .assign(mktcap=lambda x: x["mktcap"].replace(0, np.nan))
)
```

The next variable we frequently use is the one-month *lagged* market capitalization. Lagged market capitalization is typically used to compute value-weighted portfolio returns, as we demonstrate in a later chapter. The most simple and consistent way to add a column with lagged market cap values is to add one month to each observation and then join the information to our monthly CRSP data.

```
mktcap_lag = (crsp_monthly
  .assign(
    month=lambda x: x["month"]+pd.DateOffset(months=1),
    mktcap_lag=lambda x: x["mktcap"]
  )
  .get(["permno", "month", "mktcap_lag"])
)

crsp_monthly = (crsp_monthly
  .merge(mktcap_lag, how="left", on=["permno", "month"])
)
```

Next, we transform primary listing exchange codes to explicit exchange names.

```
def assign_exchange(primaryexch):
    if primaryexch == "N":
        return "NYSE"
    elif primaryexch == "A":
        return "AMEX"
    elif primaryexch == "Q":
        return "NASDAQ"
    else:
        return "Other"

crsp_monthly["exchange"] = (crsp_monthly["primaryexch"]
  .apply(assign_exchange)
)
```

Similarly, we transform industry codes to industry descriptions following Bali et al. (2016). Notice that there are also other categorizations of industries (e.g., Fama and French, 1997) that are commonly used.

```
def assign_industry(siccd):
    if 1 <= siccd <= 999:
        return "Agriculture"
    elif 1000 <= siccd <= 1499:
        return "Mining"
    elif 1500 <= siccd <= 1799:
        return "Construction"
    elif 2000 <= siccd <= 3999:
        return "Manufacturing"
    elif 4000 <= siccd <= 4899:
        return "Transportation"
    elif 4900 <= siccd <= 4999:
```

```
        return "Utilities"
    elif 5000 <= siccd <= 5199:
        return "Wholesale"
    elif 5200 <= siccd <= 5999:
        return "Retail"
    elif 6000 <= siccd <= 6799:
        return "Finance"
    elif 7000 <= siccd <= 8999:
        return "Services"
    elif 9000 <= siccd <= 9999:
        return "Public"
    else:
        return "Missing"

crsp_monthly["industry"] = (crsp_monthly["siccd"]
  .apply(assign_industry)
)
```

Next, we compute excess returns by subtracting the monthly risk-free rate provided by our Fama-French data. As we base all our analyses on the excess returns, we can drop the risk-free rate from our data frame. Note that we ensure excess returns are bounded by -1 from below as a return less than -100% makes no sense conceptually. Before we can adjust the returns, we have to connect to our database and load the data frame `factors_ff3_monthly`.

```
tidy_finance = sqlite3.connect(database="data/tidy_finance_python.sqlite")

factors_ff3_monthly = pd.read_sql_query(
  sql="SELECT month, rf FROM factors_ff3_monthly",
  con=tidy_finance,
  parse_dates={"month"}
)

crsp_monthly = (crsp_monthly
  .merge(factors_ff3_monthly, how="left", on="month")
  .assign(ret_excess=lambda x: x["ret"]-x["rf"])
  .assign(ret_excess=lambda x: x["ret_excess"].clip(lower=-1))
  .drop(columns=["rf"])
)
```

Since excess returns and market capitalization are crucial for all our analyses, we can safely exclude all observations with missing returns or market capitalization.

```
crsp_monthly = (crsp_monthly
  .dropna(subset=["ret_excess", "mktcap", "mktcap_lag"])
)
```

Finally, we store the monthly CRSP file in our database.

```
(crsp_monthly
  .to_sql(name="crsp_monthly",
          con=tidy_finance,
          if_exists="replace",
```

```
            index=False)
)
```

---

## 4.3  First Glimpse of the CRSP Sample

Before we move on to other data sources, let us look at some descriptive statistics of the CRSP sample, which is our main source for stock returns.

Figure 4.1 shows the monthly number of securities by listing exchange over time. NYSE has the longest history in the data, but NASDAQ lists a considerably large number of stocks. The number of stocks listed on AMEX decreased steadily over the last couple of decades. By the end of 2022, there were 2,778 stocks with a primary listing on NASDAQ, 1,358 on NYSE, 162 on AMEX, and only one belonged to the other category. Specifically, we use the `size()` function here to count the number of observations for each `exchange-date` group.

```
securities_per_exchange = (crsp_monthly
  .groupby(["exchange", "date"])
  .size()
  .reset_index(name="n")
)

securities_per_exchange_figure = (
  ggplot(securities_per_exchange,
         aes(x="date", y="n", color="exchange", linetype="exchange")) +
  geom_line() +
  labs(x="", y="", color="", linetype="",
       title="Monthly number of securities by listing exchange") +
  scale_x_datetime(date_breaks="10 years", date_labels="%Y") +
  scale_y_continuous(labels=comma_format())
)
securities_per_exchange_figure.draw()
```

Next, we look at the aggregate market capitalization grouped by the respective listing exchanges in Figure 4.2. To ensure that we look at meaningful data that is comparable over time, we adjust the nominal values for inflation. In fact, we can use the tables that are already in our database to calculate aggregate market caps by listing exchange. All values in Figure 4.2 are in terms of the end of 2022 USD to ensure intertemporal comparability. NYSE-listed stocks have by far the largest market capitalization, followed by NASDAQ-listed stocks.

```
cpi_monthly = pd.read_sql_query(
  sql="SELECT * FROM cpi_monthly",
  con=tidy_finance,
  parse_dates={"month"}
)

market_cap_per_exchange = (crsp_monthly
  .merge(cpi_monthly, how="left", on="month")
  .groupby(["month", "exchange"])
```

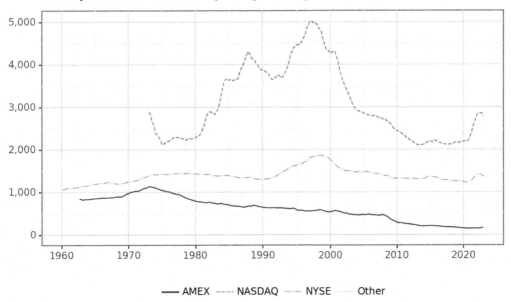

Figure 4.1: The figure shows the monthly number of stocks in the CRSP sample listed at each of the US exchanges.

```
  .apply(
    lambda group: pd.Series({
      "mktcap": group["mktcap"].sum()/group["cpi"].mean()
    })
  )
  .reset_index()
)

market_cap_per_exchange_figure = (
  ggplot(market_cap_per_exchange,
         aes(x="month", y="mktcap/1000",
             color="exchange", linetype="exchange")) +
  geom_line() +
  labs(x="", y="", color="", linetype="",
       title=("Monthly market cap by listing exchange "
              "in billions of Dec 2022 USD")) +
  scale_x_datetime(date_breaks="10 years", date_labels="%Y") +
  scale_y_continuous(labels=comma_format())
)
market_cap_per_exchange_figure.draw()
```

Next, we look at the same descriptive statistics by industry. Figure 4.3 plots the number of stocks in the sample for each of the SIC industry classifiers. For most of the sample period, the largest share of stocks is in manufacturing, albeit the number peaked somewhere in the 90s. The number of firms associated with public administration seems to be the only

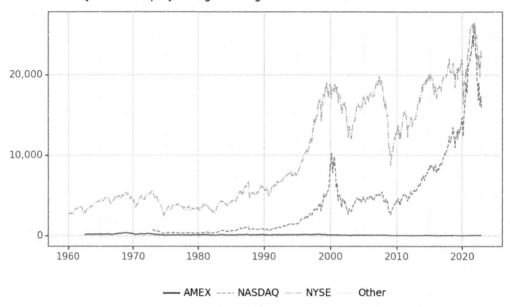

Figure 4.2: The figure shows the monthly market capitalization by listing exchange. Market capitalization is measured in billion USD, adjusted for consumer price index changes such that the values on the horizontal axis reflect the buying power of billion USD in December 2022.

category on the rise in recent years, even surpassing manufacturing at the end of our sample period.

```
securities_per_industry = (crsp_monthly
  .groupby(["industry", "date"])
  .size()
  .reset_index(name="n")
)


linetypes = ["-", "--", "-.", ":"]
n_industries = securities_per_industry["industry"].nunique()

securities_per_industry_figure = (
  ggplot(securities_per_industry,
         aes(x="date", y="n", color="industry", linetype="industry")) +
  geom_line() +
  labs(x="", y="", color="", linetype="",
       title="Monthly number of securities by industry") +
  scale_x_datetime(date_breaks="10 years", date_labels="%Y") +
  scale_y_continuous(labels=comma_format()) +
  scale_linetype_manual(
    values=[linetypes[l % len(linetypes)] for l in range(n_industries)]
  )
```

```
)
securities_per_industry_figure.draw()
```

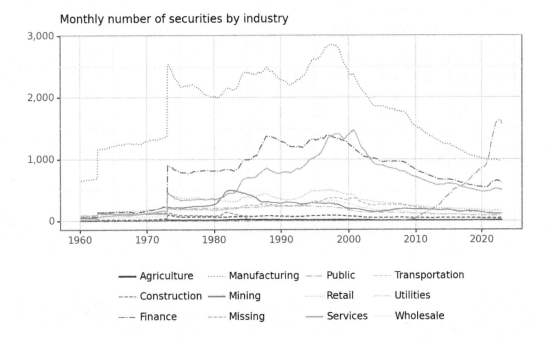

Figure 4.3: The figure shows the monthly number of stocks in the CRSP sample associated with different industries.

We also compute the market cap of all stocks belonging to the respective industries and show the evolution over time in Figure 4.4. All values are again in terms of billions of end of 2022 USD. At all points in time, manufacturing firms comprise of the largest portion of market capitalization. Toward the end of the sample, however, financial firms and services begin to make up a substantial portion of the market cap.

```
market_cap_per_industry = (crsp_monthly
  .merge(cpi_monthly, how="left", on="month")
  .groupby(["month", "industry"])
  .apply(
    lambda group: pd.Series({
      "mktcap": (group["mktcap"].sum()/group["cpi"].mean())
    })
  )
  .reset_index()
)

market_cap_per_industry_figure = (
  ggplot(market_cap_per_industry,
         aes(x="month", y="mktcap/1000",
             color="industry", linetype="industry")) +
  geom_line() +
  labs(x="", y="", color="", linetype="",
```

```
      title="Monthly market cap by industry in billions of Dec 2022 USD") +
  scale_x_datetime(date_breaks="10 years", date_labels="%Y") +
  scale_y_continuous(labels=comma_format()) +
  scale_linetype_manual(
    values=[linetypes[l % len(linetypes)] for l in range(n_industries)]
  )
)
market_cap_per_industry_figure.draw()
```

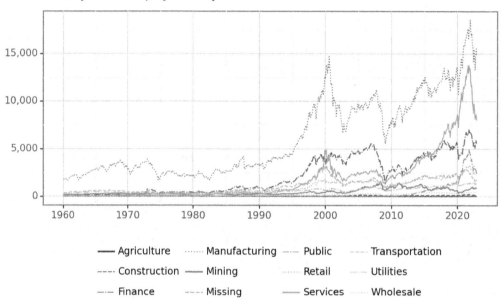

Figure 4.4: The figure shows the total Market capitalization in billion USD, adjusted for consumer price index changes such that the values on the y-axis reflect the buying power of billion USD in December 2022.

## 4.4 Daily CRSP Data

Before we turn to accounting data, we provide a proposal for downloading daily CRSP data. While the monthly data from above typically fit into your memory and can be downloaded in a meaningful amount of time, this is usually not true for daily return data. The daily CRSP data file is substantially larger than monthly data and can exceed 20GB. This has two important implications: You cannot hold all the daily return data in your memory (i.e., it is not possible to copy the entire data set to your local database), and, in our experience, the download usually crashes (or never stops) from time to time because it is too much data for the WRDS cloud to prepare and send to your Python session.

There is a solution to this challenge. As with many *big data* problems, you can split up the big task into several smaller tasks that are easier to handle. That is, instead of downloading data about all stocks at once, download the data in small batches of stocks consecutively. Such operations can be implemented in `for`-loops, where we download, prepare, and store the data for a small number of stocks in each iteration. This operation might take around five minutes, depending on your internet connection. To keep track of the progress, we create ad-hoc progress updates using `print()`. Notice that we also use the function `to_sql()` here with the option to append the new data to an existing table when we process the second and all following batches. As for the monthly CRSP data, there is no need to adjust for delisting returns in the daily CRSP data since July 2022.

```python
factors_ff3_daily = pd.read_sql(
  sql="SELECT * FROM factors_ff3_daily",
  con=tidy_finance,
  parse_dates={"date"}
)


permnos = list(crsp_monthly["permno"].unique().astype(str))

batch_size = 500
batches = np.ceil(len(permnos)/batch_size).astype(int)

for j in range(1, batches+1):

  permno_batch = permnos[
    ((j-1)*batch_size):(min(j*batch_size, len(permnos)))
  ]

  permno_batch_formatted = (
    ", ".join(f"'{permno}'" for permno in permno_batch)
  )
  permno_string = f"({permno_batch_formatted})"

  crsp_daily_sub_query = (
    "SELECT permno, dlycaldt AS date, dlyret AS ret "
      "FROM crsp.dsf_v2 "
    f"WHERE permno IN {permno_string} "
        f"AND dlycaldt BETWEEN '{start_date}' AND '{end_date}'"
  )

  crsp_daily_sub = (pd.read_sql_query(
      sql=crsp_daily_sub_query,
      con=wrds,
      dtype={"permno": int},
      parse_dates={"date"}
    )
    .dropna()
  )

  if not crsp_daily_sub.empty:

      crsp_daily_sub = (crsp_daily_sub
        .assign(
```

```
      month = lambda x:
        x["date"].dt.to_period("M").dt.to_timestamp()
    )
    .merge(factors_ff3_daily[["date", "rf"]],
          on="date", how="left")
    .assign(
      ret_excess = lambda x:
        ((x["ret"] - x["rf"]).clip(lower=-1))
    )
    .get(["permno", "date", "month", "ret_excess"])
  )

  if j == 1:
    if_exists_string = "replace"
  else:
    if_exists_string = "append"

  crsp_daily_sub.to_sql(
    name="crsp_daily",
    con=tidy_finance,
    if_exists=if_exists_string,
    index=False
  )

print(f"Batch {j} out of {batches} done ({(j/batches)*100:.2f}%)\n")
```

Eventually, we end up with more than 71 million rows of daily return data. Note that we only store the identifying information that we actually need, namely `permno`, `date`, and `month` alongside the excess returns. We thus ensure that our local database contains only the data that we actually use.

---

## 4.5   Preparing Compustat Data

Firm accounting data are an important source of information that we use in portfolio analyses in subsequent chapters. The commonly used source for firm financial information is Compustat provided by S&P Global Market Intelligence,[6] which is a global data vendor that provides financial, statistical, and market information on active and inactive companies throughout the world. For US and Canadian companies, annual history is available back to 1950 and quarterly as well as monthly histories date back to 1962.

To access Compustat data, we can again tap WRDS, which hosts the `funda` table that contains annual firm-level information on North American companies. We follow the typical filter conventions and pull only data that we actually need: (i) we get only records in industrial data format, (ii) in the standard format (i.e., consolidated information in standard presentation), and (iii) only data in the desired time window.

---

[6]https://www.spglobal.com/marketintelligence/en/

```
compustat_query = (
  "SELECT gvkey, datadate, seq, ceq, at, lt, txditc, txdb, itcb, pstkrv, "
      "pstkl, pstk, capx, oancf, sale, cogs, xint, xsga "
    "FROM comp.funda "
    "WHERE indfmt = 'INDL' "
        "AND datafmt = 'STD' "
        "AND consol = 'C' "
        f"AND datadate BETWEEN '{start_date}' AND '{end_date}'"
)

compustat = pd.read_sql_query(
  sql=compustat_query,
  con=wrds,
  dtype={"gvkey": str},
  parse_dates={"datadate"}
)
```

Next, we calculate the book value of preferred stock and equity `be` and the operating profitability `op` inspired by the variable definitions in Kenneth French's data library.[7] Note that we set negative or zero equity to missing, which is a common practice when working with book-to-market ratios (see Fama and French, 1992, for details).

```
compustat = (compustat
  .assign(
    be=lambda x:
      (x["seq"].combine_first(x["ceq"]+x["pstk"])
       .combine_first(x["at"]-x["lt"])+
       x["txditc"].combine_first(x["txdb"]+x["itcb"]).fillna(0)-
       x["pstkrv"].combine_first(x["pstkl"])
       .combine_first(x["pstk"]).fillna(0))
  )
  .assign(
    be=lambda x: x["be"].apply(lambda y: np.nan if y <= 0 else y)
  )
  .assign(
    op=lambda x:
      ((x["sale"]-x["cogs"].fillna(0)-
        x["xsga"].fillna(0)-x["xint"].fillna(0))/x["be"])
  )
)
```

We keep only the last available information for each firm-year group (by using the `tail(1)` pandas function for each group). Note that `datadate` defines the time the corresponding financial data refers to (e.g., annual report as of December 31, 2022). Therefore, `datadate` is not the date when data was made available to the public. Check out the Exercises for more insights into the peculiarities of `datadate`.

```
compustat = (compustat
  .assign(year=lambda x: pd.DatetimeIndex(x["datadate"]).year)
  .sort_values("datadate")
```

---

[7]https://mba.tuck.dartmouth.edu/pages/faculty/ken.french/data_library.html

```
    .groupby(["gvkey", "year"])
    .tail(1)
    .reset_index()
)
```

We also compute the investment ratio (`inv`) according to Kenneth French's variable definitions as the change in total assets from one fiscal year to another. Note that we again use the approach using joins as introduced with the CRSP data above to construct lagged assets.

```
compustat_lag = (compustat
  .get(["gvkey", "year", "at"])
  .assign(year=lambda x: x["year"]+1)
  .rename(columns={"at": "at_lag"})
)

compustat = (compustat
  .merge(compustat_lag, how="left", on=["gvkey", "year"])
  .assign(inv=lambda x: x["at"]/x["at_lag"]-1)
  .assign(inv=lambda x: np.where(x["at_lag"] <= 0, np.nan, x["inv"]))
)
```

With the last step, we are already done preparing the firm fundamentals. Thus, we can store them in our local database.

```
(compustat
  .to_sql(name="compustat",
          con=tidy_finance,
          if_exists="replace",
          index=False)
)
```

## 4.6   Merging CRSP with Compustat

Unfortunately, CRSP and Compustat use different keys to identify stocks and firms. CRSP uses `permno` for stocks, while Compustat uses `gvkey` to identify firms. Fortunately, a curated matching table on WRDS allows us to merge CRSP and Compustat, so we create a connection to the *CRSP-Compustat Merged* table (provided by CRSP). The linking table contains links between CRSP and Compustat identifiers from various approaches. However, we need to make sure that we keep only relevant and correct links, again following the description outlined in Bali et al. (2016). Note also that currently active links have no end date, so we just enter the current date via the SQL verb `CURRENT_DATE`.

```
ccmxpf_linktable_query = (
  "SELECT lpermno AS permno, gvkey, linkdt, "
      "COALESCE(linkenddt, CURRENT_DATE) AS linkenddt "
    "FROM crsp.ccmxpf_linktable "
    "WHERE linktype IN ('LU', 'LC') "
        "AND linkprim IN ('P', 'C') "
```

```
              "AND usedflag = 1"
)

ccmxpf_linktable = pd.read_sql_query(
  sql=ccmxpf_linktable_query,
  con=wrds,
  dtype={"permno": int, "gvkey": str},
  parse_dates={"linkdt", "linkenddt"}
)
```

We use these links to create a new table with a mapping between stock identifier, firm identifier, and month. We then add these links to the Compustat `gvkey` to our monthly stock data.

```
ccm_links = (crsp_monthly
  .merge(ccmxpf_linktable, how="inner", on="permno")
  .query("~gvkey.isnull() & (date >= linkdt) & (date <= linkenddt)")
  .get(["permno", "gvkey", "date"])
)

crsp_monthly = (crsp_monthly
  .merge(ccm_links, how="left", on=["permno", "date"])
)
```

As the last step, we update the previously prepared monthly CRSP file with the linking information in our local database.

```
(crsp_monthly
  .to_sql(name="crsp_monthly",
          con=tidy_finance,
          if_exists="replace",
          index=False)
)
```

Before we close this chapter, let us look at an interesting descriptive statistic of our data. As the book value of equity plays a crucial role in many asset pricing applications, it is interesting to know for how many of our stocks this information is available. Hence, Figure 4.5 plots the share of securities with book equity values for each exchange. It turns out that the coverage is pretty bad for AMEX- and NYSE-listed stocks in the 1960s but hovers around 80 percent for all periods thereafter. We can ignore the erratic coverage of securities that belong to the other category since there is only a handful of them anyway in our sample.

```
share_with_be = (crsp_monthly
  .assign(year=lambda x: pd.DatetimeIndex(x["month"]).year)
  .sort_values("date")
  .groupby(["permno", "year"])
  .tail(1)
  .reset_index()
  .merge(compustat, how="left", on=["gvkey", "year"])
  .groupby(["exchange", "year"])
  .apply(
    lambda x: pd.Series({
```

```
  "share": x["permno"][~x["be"].isnull()].nunique()/x["permno"].nunique()
    })
  )
  .reset_index()
)

share_with_be_figure = (
  ggplot(share_with_be,
         aes(x="year", y="share", color="exchange", linetype="exchange")) +
  geom_line() +
  labs(x="", y="", color="", linetype="",
       title="Share of securities with book equity values by exchange") +
  scale_y_continuous(labels=percent_format()) +
  coord_cartesian(ylim=(0, 1))
)
share_with_be_figure.draw()
```

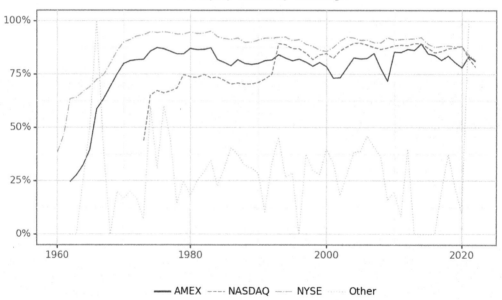

Figure 4.5: The figure shows the end-of-year share of securities with book equity values by listing exchange.

## 4.7    Exercises

1.  Compute `mkt_cap_lag` using `lag(mktcap)` rather than using joins as above. Filter out all the rows where the lag-based market capitalization measure is different from the one we computed above. Why are the two measures different?

2. Plot the average market capitalization of firms for each exchange and industry, respectively, over time. What do you find?

3. In the `compustat` table, `datadate` refers to the date to which the fiscal year of a corresponding firm refers. Count the number of observations in Compustat by `month` of this date variable. What do you find? What does the finding suggest about pooling observations with the same fiscal year?

4. Go back to the original Compustat and extract rows where the same firm has multiple rows for the same fiscal year. What is the reason for these observations?

5. Keep the last observation of `crsp_monthly` by year and join it with the `compustat` table. Create the following plots: (i) aggregate book equity by exchange over time and (ii) aggregate annual book equity by industry over time. Do you notice any different patterns to the corresponding plots based on market capitalization?

6. Repeat the analysis of market capitalization for book equity, which we computed from the Compustat data. Then, use the matched sample to plot book equity against market capitalization. How are these two variables related?

# 5

# *TRACE and FISD*

In this chapter, we dive into the US corporate bond market. Bond markets are far more diverse than stock markets, as most issuers have multiple bonds outstanding simultaneously with potentially very different indentures. This market segment is exciting due to its size (roughly ten trillion USD outstanding), heterogeneity of issuers (as opposed to government bonds), market structure (mostly over-the-counter trades), and data availability. We introduce how to use bond characteristics from FISD and trade reports from TRACE and provide code to download and clean TRACE in Python.

Many researchers study liquidity in the US corporate bond market, with notable contributions from Bessembinder et al. (2006), Edwards et al. (2007), and O'Hara and Zhou (2021), among many others. We do not cover bond returns here, but you can compute them from TRACE data. Instead, we refer to studies on the topic such as Bessembinder et al. (2008), Bai et al. (2019), and Kelly et al. (2021) and a survey by Huang and Shi (2021).

This chapter also draws on the resources provided by the project Open Source Bond Asset Pricing[1] and their related publication, i.e., Dickerson et al. (2023). We encourage you to visit their website to check out the additional resources they provide. Moreover, WRDS provides bond returns computed from TRACE data at a monthly frequency.

The current chapter relies on the following set of Python packages.

```python
import pandas as pd
import numpy as np
import sqlite3
import httpimport

from plotnine import *
from sqlalchemy import create_engine
from mizani.breaks import date_breaks
from mizani.formatters import date_format, comma_format
```

Compared to previous chapters, we load `httpimport` (Torakis, 2023) to source code provided in the public Gist.[2] Note that you should be careful with loading anything from the web via this method, and it is highly discouraged to use any unsecured "HTTP" links. Also, you might encounter a problem when using this from a corporate computer that prevents downloading data through a firewall.

---

[1]https://openbondassetpricing.com/
[2]https://docs.github.com/en/get-started/writing-on-github/editing-and-sharing-content-with-gists/creating-gists

## 5.1   Bond Data from WRDS

Both bond databases we need are available on WRDS[3] to which we establish the PostgreSQL connection described in Chapter 4. Additionally, we connect to our local SQLite-database to store the data we download.

```
import os
from dotenv import load_dotenv
load_dotenv()

connection_string = (
  "postgresql+psycopg2://"
  f"{os.getenv('WRDS_USER')}:{os.getenv('WRDS_PASSWORD')}"
  "@wrds-pgdata.wharton.upenn.edu:9737/wrds"
)

wrds = create_engine(connection_string, pool_pre_ping=True)

tidy_finance = sqlite3.connect(database="data/tidy_finance_python.sqlite")
```

## 5.2   Mergent FISD

For research on US corporate bonds, the Mergent Fixed Income Securities Database (FISD) is the primary resource for bond characteristics. There is a detailed manual[4] on WRDS, so we only cover the necessary subjects here. FISD data comes in two main variants, namely, centered on issuers or issues. In either case, the most useful identifiers are CUSIPs.[5] 9-digit CUSIPs identify securities issued by issuers. The issuers can be identified from the first six digits of a security CUSIP, which is also called a 6-digit CUSIP. Both stocks and bonds have CUSIPs. This connection would, in principle, allow matching them easily, but due to changing issuer details, this approach only yields small coverage.

We use the issue-centered version of FISD to identify the subset of US corporate bonds that meet the standard criteria (Bessembinder et al., 2006). The WRDS table `fisd_mergedissue` contains most of the information we need on a 9-digit CUSIP level. Due to the diversity of corporate bonds, details in the indenture vary significantly. We focus on common bonds that make up the majority of trading volume in this market without diverging too much in indentures.

The following chunk connects to the data and selects the bond sample to remove certain bond types that are less commonly used (see, e.g., Dick-Nielsen et al., 2012; O'Hara and Zhou, 2021, among many others). In particular, we use the filters listed below. Note that we also treat missing values in these flags.

---

[3] https://wrds-www.wharton.upenn.edu/

[4] https://wrds-www.wharton.upenn.edu/documents/1364/FixedIncome_Securities_Master_Database_User_Guide_v4.pdf

[5] https://www.cusip.com/index.html

1. Keep only senior bonds (`security_level = 'SEN'`).
2. Exclude bonds which are secured lease obligations (`slob = 'N' OR slob IS NULL`).
3. Exclude secured bonds (`security_pledge IS NULL`).
4. Exclude asset-backed bonds (`asset_backed = 'N' OR asset_backed IS NULL`).
5. Exclude defeased bonds (`(defeased = 'N' OR defeased IS NULL) AND defeased_date IS NULL`).
6. Keep only the bond types US Corporate Debentures (`'CDEB'`), US Corporate Medium Term Notes (`'CMTN'`), US Corporate Zero Coupon Notes and Bonds (`'CMTZ', 'CZ'`), and US Corporate Bank Note (`'USBN'`).
7. Exclude bonds that are payable in kind (`(pay_in_kind != 'Y' OR pay_in_kind IS NULL) AND pay_in_kind_exp_date IS NULL`).
8. Exclude foreign (`yankee == "N" OR is.na(yankee)`) and Canadian issuers (`canadian = 'N' OR canadian IS NULL`).
9. Exclude bonds denominated in foreign currency (`foreign_currency = 'N'`).
10. Keep only fixed (F) and zero (Z) coupon bonds with additional requirements of `fix_frequency IS NULL`, `coupon_change_indicator = 'N'` and annual, semi-annual, quarterly, or monthly interest frequencies.
11. Exclude bonds that were issued under SEC Rule 144A (`rule_144a = 'N'`).
12. Exlcude privately placed bonds (`private_placement = 'N' OR private_placement IS NULL`).
13. Exclude defaulted bonds (`defaulted = 'N' AND filing_date IS NULL AND settlement IS NULL`).
14. Exclude convertible (`convertible = 'N'`), putable (`putable = 'N' OR putable IS NULL`), exchangeable (`exchangeable = 'N' OR exchangeable IS NULL`), perpetual (`perpetual = 'N'`), or preferred bonds (`preferred_security = 'N' OR preferred_security IS NULL`).
15. Exclude unit deal bonds (`(unit_deal = 'N' OR unit_deal IS NULL)`).

```
fisd_query = (
  "SELECT complete_cusip, maturity, offering_amt, offering_date, "
        "dated_date, interest_frequency, coupon, last_interest_date, "
        "issue_id, issuer_id "
   "FROM fisd.fisd_mergedissue "
   "WHERE security_level = 'SEN' "
        "AND (slob = 'N' OR slob IS NULL) "
        "AND security_pledge IS NULL "
        "AND (asset_backed = 'N' OR asset_backed IS NULL) "
        "AND (defeased = 'N' OR defeased IS NULL) "
        "AND defeased_date IS NULL "
        "AND bond_type IN ('CDEB', 'CMTN', 'CMTZ', 'CZ', 'USBN') "
        "AND (pay_in_kind != 'Y' OR pay_in_kind IS NULL) "
        "AND pay_in_kind_exp_date IS NULL "
        "AND (yankee = 'N' OR yankee IS NULL) "
        "AND (canadian = 'N' OR canadian IS NULL) "
        "AND foreign_currency = 'N' "
        "AND coupon_type IN ('F', 'Z') "
        "AND fix_frequency IS NULL "
        "AND coupon_change_indicator = 'N' "
        "AND interest_frequency IN ('0', '1', '2', '4', '12') "
```

```
            "AND rule_144a = 'N' "
            "AND (private_placement = 'N' OR private_placement IS NULL) "
            "AND defaulted = 'N' "
            "AND filing_date IS NULL "
            "AND settlement IS NULL "
            "AND convertible = 'N' "
            "AND exchange IS NULL "
            "AND (putable = 'N' OR putable IS NULL) "
            "AND (unit_deal = 'N' OR unit_deal IS NULL) "
            "AND (exchangeable = 'N' OR exchangeable IS NULL) "
            "AND perpetual = 'N' "
            "AND (preferred_security = 'N' OR preferred_security IS NULL)"
)

fisd = pd.read_sql_query(
  sql=fisd_query,
  con=wrds,
  dtype={"complete_cusip": str, "interest_frequency": str,
          "issue_id": int, "issuer_id": int},
  parse_dates={"maturity", "offering_date",
                "dated_date", "last_interest_date"}
)
```

We also pull issuer information from `fisd_mergedissuer` regarding the industry and country of the firm that issued a particular bond. Then, we filter to include only US-domiciled firms' bonds. We match the data by `issuer_id`.

```
fisd_issuer_query = (
  "SELECT issuer_id, sic_code, country_domicile "
    "FROM fisd.fisd_mergedissuer"
)

fisd_issuer = pd.read_sql_query(
  sql=fisd_issuer_query,
  con=wrds,
  dtype={"issuer_id": int, "sic_code": str, "country_domicile": str}
)

fisd = (fisd
  .merge(fisd_issuer, how="inner", on="issuer_id")
  .query("country_domicile == 'USA'")
  .drop(columns="country_domicile")
)
```

Finally, we save the bond characteristics to our local database. This selection of bonds also constitutes the sample for which we will collect trade reports from TRACE below.

```
(fisd
  .to_sql(name="fisd",
          con=tidy_finance,
          if_exists="replace",
```

```
        index=False)
)
```

The FISD database also contains other data. The issue-based file contains information on covenants, i.e., restrictions included in bond indentures to limit specific actions by firms (e.g., Handler et al., 2021). The FISD redemption database also contains information on callable bonds. Moreover, FISD also provides information on bond ratings. We do not need either here.

## 5.3 TRACE

The Financial Industry Regulatory Authority (FINRA) provides the Trade Reporting and Compliance Engine (TRACE). In TRACE, dealers that trade corporate bonds must report such trades individually. Hence, we observe trade messages in TRACE that contain information on the bond traded, the trade time, price, and volume. TRACE comes in two variants: standard and enhanced TRACE. We show how to download and clean enhanced TRACE as it contains uncapped volume, a crucial quantity missing in the standard distribution. Moreover, enhanced TRACE also provides information on the respective parties' roles and the direction of the trade report. These items become essential in cleaning the messages.

Why do we repeatedly talk about cleaning TRACE? Trade messages are submitted within a short time window after a trade is executed (less than 15 minutes). These messages can contain errors, and the reporters subsequently correct them or they cancel a trade altogether. The cleaning needs are described by Dick-Nielsen (2009) in detail, and Dick-Nielsen (2014) shows how to clean the enhanced TRACE data using SAS. We do not go into the cleaning steps here, since the code is lengthy and is not our focus here. However, downloading and cleaning enhanced TRACE data is straightforward with our setup.

We store code for cleaning enhanced TRACE with Python on the following GitHub Gist.[6] Appendix D also contains the code for reference. We only need to source the code from the Gist, which we can do with the code below using `httpimport`. In the chunk, we explicitly load the necessary function interpreting the Gist as a module (i.e., you could also use it as a module and precede the function calls with the module's name). Alternatively, you can also go to the Gist, download it, and manually execute it. The `clean_enhanced_trace()` function takes a vector of CUSIPs, a connection to WRDS explained in Chapter 4, and a start and end date, respectively.

```
gist_url = (
  "https://gist.githubusercontent.com/patrick-weiss/"
  "86ddef6de978fbdfb22609a7840b5d8b/raw/"
  "8fbcc6c6f40f537cd3cd37368be4487d73569c6b/"
)
```

```
with httpimport.remote_repo(gist_url):
  from clean_enhanced_TRACE_python import clean_enhanced_trace
```

The TRACE database is considerably large. Therefore, we only download subsets of data at once. Specifying too many CUSIPs over a long time horizon will result in very long download

---

[6]https://gist.githubusercontent.com/patrick-weiss/86ddef6de978fbdfb22609a7840b5d8b

times and a potential failure due to the size of the request to WRDS. The size limit depends on many parameters, and we cannot give you a guideline here. For the applications in this book, we need data around the Paris Agreement in December 2015 and download the data in sets of 1000 bonds, which we define below.

```python
cusips = list(fisd["complete_cusip"].unique())
batch_size = 1000
batches = np.ceil(len(cusips)/batch_size).astype(int)
```

Finally, we run a loop in the same style as in Chapter 4 where we download daily returns from CRSP. For each of the CUSIP sets defined above, we call the cleaning function and save the resulting output. We add new data to the existing table for batch two and all following batches.

```python
for j in range(1, batches + 1):

    cusip_batch = cusips[
        ((j-1)*batch_size):(min(j*batch_size, len(cusips)))
    ]

    cusip_batch_formatted = ", ".join(f"'{cusip}'" for cusip in cusip_batch)
    cusip_string = f"({permno_batch_formatted})"

    trace_enhanced_sub = clean_enhanced_trace(
        cusips=cusip_string,
        connection=wrds,
        start_date="'01/01/2014'",
        end_date="'11/30/2016'"
    )

    if not trace_enhanced_sub.empty:
        if j == 1:
            if_exists_string = "replace"
        else:
            if_exists_string = "append"

        trace_enhanced_sub.to_sql(
            name="trace_enhanced",
            con=tidy_finance,
            if_exists=if_exists_string,
            index=False
        )

    print(f"Batch {j} out of {batches} done ({(j/batches)*100:.2f}%)\n")
```

## 5.4   Insights into Corporate Bonds

While many news outlets readily provide information on stocks and the underlying firms, corporate bonds are not covered frequently. Additionally, the TRACE database contains trade-level information, potentially new to students. Therefore, we provide you with some insights by showing some summary statistics.

We start by looking into the number of bonds outstanding over time and compare it to the number of bonds traded in our sample. First, we compute the number of bonds outstanding for each quarter around the Paris Agreement from 2014 to 2016.

```python
dates = pd.date_range(start="2014-01-01", end="2016-11-30", freq="Q")

bonds_outstanding = (pd.DataFrame({"date": dates})
  .merge(fisd[["complete_cusip"]], how="cross")
  .merge(fisd[["complete_cusip", "offering_date", "maturity"]],
         on="complete_cusip", how="left")
  .assign(offering_date=lambda x: x["offering_date"].dt.floor("D"),
          maturity=lambda x: x["maturity"].dt.floor("D"))
  .query("date >= offering_date & date <= maturity")
  .groupby("date")
  .size()
  .reset_index(name="count")
  .assign(type="Outstanding")
)
```

Next, we look at the bonds traded each quarter in the same period. Notice that we load the complete trace table from our database, as we only have a single part of it in the environment from the download loop above.

```python
trace_enhanced = pd.read_sql_query(
  sql=("SELECT cusip_id, trd_exctn_dt, rptd_pr, entrd_vol_qt, yld_pt "
       "FROM trace_enhanced"),
  con=tidy_finance,
  parse_dates={"trd_exctn_dt"}
)

bonds_traded = (trace_enhanced
  .assign(
    date=lambda x: (
      (x["trd_exctn_dt"]-pd.offsets.MonthBegin(1))
        .dt.to_period("Q").dt.start_time
    )
  )
  .groupby("date")
  .aggregate(count=("cusip_id", "nunique"))
  .reset_index()
  .assign(type="Traded")
)
```

Finally, we plot the two time series in Figure 5.1.

```python
bonds_combined = pd.concat(
  [bonds_outstanding, bonds_traded], ignore_index=True
)

bonds_figure = (
```

```
ggplot(bonds_combined,
       aes(x="date", y="count", color="type", linetype="type")) +
  geom_line() +
  labs(x="", y="", color="", linetype="",
       title="Number of bonds outstanding and traded each quarter") +
  scale_x_datetime(breaks=date_breaks("1 year"), labels=date_format("%Y")) +
  scale_y_continuous(labels=comma_format())
)
bonds_figure.draw()
```

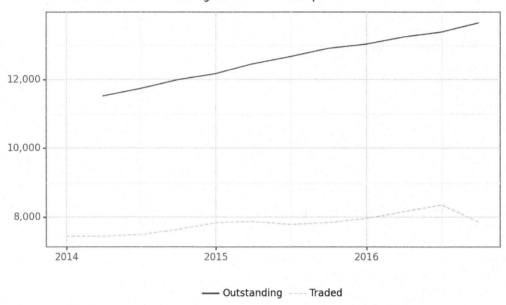

Figure 5.1: Number of corporate bonds outstanding each quarter as reported by Mergent FISD and the number of traded bonds from enhanced TRACE between 2014 and end of 2016.

We see that the number of bonds outstanding increases steadily between 2014 and 2016. During our sample period of trade data, we see that the fraction of bonds trading each quarter is roughly 60 percent. The relatively small number of traded bonds means that many bonds do not trade through an entire quarter. This lack of trading activity illustrates the generally low level of liquidity in the corporate bond market, where it can be hard to trade specific bonds. Does this lack of liquidity mean that corporate bond markets are irrelevant in terms of their size? With over 7,500 traded bonds each quarter, it is hard to say that the market is small. However, let us also investigate the characteristics of issued corporate bonds. In particular, we consider maturity (in years), coupon, and offering amount (in million USD).

```
average_characteristics = (fisd
  .assign(
    maturity=lambda x: (x["maturity"] - x["offering_date"]).dt.days/365,
    offering_amt=lambda x: x["offering_amt"]/10**3
```

```
  )
  .melt(var_name="measure",
       value_vars=["maturity", "coupon", "offering_amt"],
       value_name="value")
  .dropna()
  .groupby("measure")["value"]
  .describe(percentiles=[.05, .50, .95])
  .drop(columns="count")
)
average_characteristics.round(2)
```

| measure | mean | std | min | 5% | 50% | 95% | max |
|---|---|---|---|---|---|---|---|
| coupon | 2.78 | 3.59 | 0.00 | 0.00 | 0.00 | 9.00 | 39.00 |
| maturity | 5.96 | 7.20 | -6.24 | 1.04 | 4.02 | 24.97 | 180.13 |
| offering_amt | 145.92 | 386.22 | 0.00 | 0.24 | 3.74 | 763.60 | 15000.00 |

We see that the average bond in our sample period has an offering amount of over 357 million USD with a median of 200 million USD, which both cannot be considered small. The average bond has a maturity of ten years and pays around 6 percent in coupons.

Finally, let us compute some summary statistics for the trades in this market. To this end, we show a summary based on aggregate information daily. In particular, we consider the trade size (in million USD) and the number of trades.

```
average_trade_size = (trace_enhanced
  .groupby("trd_exctn_dt")
  .aggregate(
    trade_size=("entrd_vol_qt", lambda x: (
      sum(x*trace_enhanced.loc[x.index, "rptd_pr"]/100)/10**6)
    ),
    trade_number=("trd_exctn_dt", "size")
  )
  .reset_index()
  .melt(id_vars=["trd_exctn_dt"], var_name="measure",
       value_vars=["trade_size", "trade_number"], value_name="value")
  .groupby("measure")["value"]
  .describe(percentiles=[.05, .50, .95])
  .drop(columns="count")
)
average_trade_size.round(0)
```

| measure | mean | std | min | 5% | 50% | 95% | max |
|---|---|---|---|---|---|---|---|
| trade_number | 25914.0 | 5444.0 | 439.0 | 17844.0 | 26051.0 | 34383.0 | 40839.0 |
| trade_size | 12968.0 | 3577.0 | 17.0 | 6128.0 | 13421.0 | 17850.0 | 21312.0 |

On average, nearly 26 billion USD of corporate bonds are traded daily in nearly 13,000 transactions. We can, hence, conclude that the corporate bond market is indeed significant in terms of trading volume and activity.

## 5.5   Exercises

1. Compute the amount outstanding across all bonds over time. Make sure to subtract all matured bonds. How would you describe the resulting plot?

2. Compute the number of days each bond is traded (accounting for the bonds' maturities and issuances). Start by looking at the number of bonds traded each day in a graph similar to the one above. How many bonds trade on more than 75 percent of trading days?

3. WRDS provides more information from Mergent FISD such as ratings in the table `fisd_ratings`. Download the ratings table and plot the distribution of ratings for the different rating providers. How would you map the different providers to a common numeric rating scale?

# 6

## *Other Data Providers*

In the previous chapters, we introduced many ways to get financial data that researchers regularly use. We showed how to load data into Python from Yahoo!Finance (using the `yfinance` package) and from Kenneth French's data library[1] (using the `pandas-datareader` package). We also presented commonly used file types, such as comma-separated or Excel files. Then, we introduced remotely connecting to WRDS and downloading data from there. However, this is only a subset of the vast amounts of data available online these days.

In this chapter, we provide an overview of common alternative data providers for which direct access via Python packages exists. Such a list requires constant adjustments because both data providers and access methods change. However, we want to emphasize two main insights. First, the number of Python packages that provide access to (financial) data is large. Too large actually to survey here exhaustively. Instead, we can only cover the tip of the iceberg. Second, Python provides the functionalities to access any form of files or data available online. Thus, even if a desired data source does not come with a well-established Python package, chances are high that data can be retrieved by establishing your own API connection (using the Python `requests` package) or by scrapping the content.

In our non-exhaustive list below, we restrict ourselves to listing data sources accessed through easy-to-use Python packages. For further inspiration on potential data sources, we recommend reading the Awesome Quant curated list of insanely awesome libraries, packages, and resources for Quants (Quantitative Finance).[2] In fact, the `pandas-datareader` package provides comprehensive access to a lot of databases, including some of those listed below.

Also, the `requests` library in Python provides a versatile and direct way to interact with APIs (Application Programming Interfaces) offered by various financial data providers. The package simplifies the process of making HTTP requests,[3] handling authentication, and parsing the received data.

Apart from the list below, we want to advertise some amazing data compiled by others. First, there is Open Source Asset Pricing[4] related to Chen and Zimmermann (2022). They provide return data for over 200 trading strategies with different time periods and specifications. The authors also provide signals and explanations of the factor construction. Moreover, in the same spirit, Global factor data[5] provides the data related to Jensen2022b. They provide return data for characteristic-managed portfolios from around the world. The database includes factors for 153 characteristics in 13 themes, using data from 93 countries. Finally, we want to mention the TAQ data providing trades and quotes data for NYSE, Nasdaq, and regional exchanges, which is available via WRDS.

---

[1] https://mba.tuck.dartmouth.edu/pages/faculty/ken.french/data_library.html
[2] https://github.com/wilsonfreitas/awesome-quant
[3] https://requests.readthedocs.io/
[4] https://www.openassetpricing.com/data
[5] https://www.jkpfactors.com

| Source | Description | Packages |
| --- | --- | --- |
| FRED | The Federal Reserve Bank of St. Louis provides more than 818,000 US and international time series from 109 sources via the FRED API. Data can be browsed online on the FRED homepage. | `fredapi`, `pandas-datareader` |
| ECB | The European Central Bank's Statistical Data Warehouse offers data on Euro area monetary policy, financial stability, and more. | `sdmx` |
| OECD | The OECD monitors trends, collects data, and forecasts economic development in various public policy areas. It's a major source of comparable statistical, economic, and social data. | `requests`, `pandas-datareader` |
| World Bank | Provides access to global statistics including World Development Indicators, International Debt Statistics, and more, available for over 200 countries. | `requests`, `pandas-datareader` |
| Eurostat | Eurostat, the EU's statistical office, provides high-quality data on Europe. | `requests`, `pandas-datareader` |
| Econdb | Offers access to economic data from over 90 official statistical agencies through a comprehensive database. | `requests`, `pandas-datareader` |
| Bloomberg | Provides data on balance sheets, income statements, cash flows, and more, with industry-specific data in various sectors. Paid subscription required. | `blpapi` |
| LSEG Eikon | Eikon offers real-time market data, news, analytics, and more. It's a paid service of LSEG. | `refinitiv-data` |
| Nasdaq Data Link | Quandl publishes alternative data, with some requiring specific subscriptions. | `nasdaqdatalink`, `pandas-datareader` |
| Simfin | Automates data collection to offer financial data freely to investors, researchers, and students. | `simfin` |
| PyAnomaly | A Python library for asset pricing research, offering various portfolio analytics tools. | `pyanomaly` |
| IEX | Operates the IEX, offering US reference and market data, including intraday pricing data. | `pyEx`, `pandas-datareader` |
| CoinMarketCap | Offers cryptocurrency information, historical prices, and exchange listings. | `coinmarketcap` |
| CoinGecko | An alternative crypto data provider for current and historical coin and exchange data. | `pycoingecko` |
| X (Twitter) | Offers limited access for academic research on Tweets. | `tweepy` |

| Source | Description | Packages |
|--------|-------------|----------|
| SEC company filings | The EDGAR database provides public access to corporate and mutual fund information filed with the SEC. | `sec-api` |
| Google Trends | Provides public access to global search volumes via Google Trends. | `pytrends` |

## 6.1 Exercises

1. Select one of the data sources in the table above and retrieve some data. Browse the homepage of the data provider or the package documentation to find inspiration on which type of data is available to you and how to download the data into your Python session.

2. Generate summary statistics of the data you retrieved and provide some useful visualization. The possibilities are endless: Maybe there is some interesting economic event you want to analyze, such as stock market responses to Twitter activity.

# Part III

# Asset Pricing

# 7

# *Beta Estimation*

In this chapter, we introduce an important concept in financial economics: The exposure of an individual stock to changes in the market portfolio. According to the Capital Asset Pricing Model (CAPM) of Sharpe (1964), Lintner (1965), and Mossin (1966), cross-sectional variation in expected asset returns should be a function of the covariance between the excess return of the asset and the excess return on the market portfolio. The regression coefficient of excess market returns on excess stock returns is usually called the market beta. We show an estimation procedure for the market betas. We do not go into details about the foundations of market beta but simply refer to any treatment of the CAPM[1] for further information. Instead, we provide details about all the functions that we use to compute the results. In particular, we leverage useful computational concepts: Rolling-window estimation and parallelization.

We use the following Python packages throughout this chapter:

```python
import pandas as pd
import numpy as np
import sqlite3
import statsmodels.formula.api as smf

from regtabletotext import prettify_result
from statsmodels.regression.rolling import RollingOLS
from plotnine import *
from mizani.breaks import date_breaks
from mizani.formatters import percent_format, date_format
from joblib import Parallel, delayed, cpu_count
from itertools import product
```

Compared to previous chapters, we introduce `statsmodels` (Seabold and Perktold, 2010) for regression analysis and for sliding-window regressions and `joblib` (Team, 2023) for parallelization.

## 7.1 Estimating Beta Using Monthly Returns

The estimation procedure is based on a rolling-window estimation, where we may use either monthly or daily returns and different window lengths. First, let us start with loading the monthly CRSP data from our SQLite database introduced in Chapter 3 and Chapter 4.

---

[1] https://en.wikipedia.org/wiki/Capital_asset_pricing_model

```
tidy_finance = sqlite3.connect(database="data/tidy_finance_python.sqlite")

crsp_monthly = (pd.read_sql_query(
    sql="SELECT permno, month, industry, ret_excess FROM crsp_monthly",
    con=tidy_finance,
    parse_dates={"month"})
  .dropna()
)

factors_ff3_monthly = pd.read_sql_query(
  sql="SELECT month, mkt_excess FROM factors_ff3_monthly",
  con=tidy_finance,
  parse_dates={"month"}
)

crsp_monthly = (crsp_monthly
  .merge(factors_ff3_monthly, how="left", on="month")
)
```

To estimate the CAPM regression coefficients

$$r_{i,t} - r_{f,t} = \alpha_i + \beta_i(r_{m,t} - r_{f,t}) + \varepsilon_{i,t}, \tag{7.1}$$

we regress stock excess returns `ret_excess` on excess returns of the market portfolio `mkt_excess`.

Python provides a simple solution to estimate (linear) models with the function `smf.ols()`. The function requires a formula as input that is specified in a compact symbolic form. An expression of the form `y ~ model` is interpreted as a specification that the response `y` is modeled by a linear predictor specified symbolically by `model`. Such a model consists of a series of terms separated by + operators. In addition to standard linear models, `smf.ols()` provides a lot of flexibility. You should check out the documentation for more information. To start, we restrict the data only to the time series of observations in CRSP that correspond to Apple's stock (i.e., to `permno` 14593 for Apple) and compute $\hat{\alpha}_i$ as well as $\hat{\beta}_i$.

```
model_beta = (smf.ols(
    formula="ret_excess ~ mkt_excess",
    data=crsp_monthly.query("permno == 14593"))
  .fit()
)
prettify_result(model_beta)
```

```
OLS Model:
ret_excess ~ mkt_excess

Coefficients:
            Estimate  Std. Error  t-Statistic  p-Value
Intercept      0.010       0.005        2.004    0.046
mkt_excess     1.389       0.111       12.467    0.000

Summary statistics:
- Number of observations: 504
- R-squared: 0.236, Adjusted R-squared: 0.235
```

```
- F-statistic: 155.435 on 1 and 502 DF, p-value: 0.000
```

`smf.ols()` returns an object of class `RegressionModel`, which contains all the information we usually care about with linear models. `prettify_result()` returns an overview of the estimated parameters. The output above indicates that Apple moves excessively with the market as the estimated $\hat{\beta}_i$ is above one ($\hat{\beta}_i \approx 1.4$).

## 7.2 Rolling-Window Estimation

After we estimated the regression coefficients on an example, we scale the estimation of $\beta_i$ to a whole different level and perform rolling-window estimations for the entire CRSP sample.

We take a total of five years of data (`window_size`) and require at least 48 months with return data to compute our betas (`min_obs`). Check out the Exercises if you want to compute beta for different time periods. We first identify firm identifiers (`permno`) for which CRSP contains sufficiently many records.

```python
window_size = 60
min_obs = 48

valid_permnos = (crsp_monthly
  .dropna()
  .groupby("permno")["permno"]
  .count()
  .reset_index(name="counts")
  .query(f"counts > {window_size}+1")
)
```

Before we proceed with the estimation, one important issue is worth emphasizing: `RollingOLS` returns the estimated parameters of a linear regression by incorporating a window of the last `window_size` rows. Whenever monthly returns are implicitly missing (which means there is simply no entry recorded, e.g., because a company was delisted and only traded publicly again later), such a fixed window size may cause outdated observations to influence the estimation results. We thus recommend making such implicit missing rows explicit.

We hence collect information about the first and last listing date of each `permno`.

```python
permno_information = (crsp_monthly
  .merge(valid_permnos, how="inner", on="permno")
  .groupby(["permno"])
  .aggregate(first_month=("month", "min"),
             last_month=("month", "max"))
  .reset_index()
)
```

To complete the missing observations in the CRSP sample, we obtain all possible **permno-month** combinations.

```
unique_permno = crsp_monthly["permno"].unique()
unique_month = factors_ff3_monthly["month"].unique()

all_combinations = pd.DataFrame(
  product(unique_permno, unique_month),
  columns=["permno", "month"]
)
```

Finally, we expand the CRSP sample and include a row (with missing excess returns) for each possible **permno-month** observation that falls within the start and end date where the respective **permno** has been publicly listed.

```
returns_monthly = (all_combinations
  .merge(crsp_monthly.get(["permno", "month", "ret_excess"]),
         how="left", on=["permno", "month"])
  .merge(permno_information, how="left", on="permno")
  .query("(month >= first_month) & (month <= last_month)")
  .drop(columns=["first_month", "last_month"])
  .merge(crsp_monthly.get(["permno", "month", "industry"]),
         how="left", on=["permno", "month"])
  .merge(factors_ff3_monthly, how="left", on="month")
)
```

The following function implements the CAPM regression for a dataframe (or a part thereof) containing at least **min_obs** observations to avoid huge fluctuations if the time series is too short. If the condition is violated (i.e., the time series is too short) the function returns a missing value.

```
def roll_capm_estimation(data, window_size, min_obs):
    """Calculate rolling CAPM estimation."""

    data = data.sort_values("month")

    result = (RollingOLS.from_formula(
      formula="ret_excess ~ mkt_excess",
      data=data,
      window=window_size,
      min_nobs=min_obs,
      missing="drop")
      .fit()
      .params.get("mkt_excess")
    )

    result.index = data.index

    return result
```

Before we approach the whole CRSP sample, let us focus on a couple of examples for well-known firms.

```
examples = pd.DataFrame({
  "permno": [14593, 10107, 93436, 17778],
```

```
  "company": ["Apple", "Microsoft", "Tesla", "Berkshire Hathaway"]
})
```

It is actually quite simple to perform the rolling-window estimation for an arbitrary number of stocks, which we visualize in the following code chunk and the resulting Figure 7.1.

```
beta_example = (returns_monthly
  .merge(examples, how="inner", on="permno")
  .groupby(["permno"])
  .apply(lambda x: x.assign(
    beta=roll_capm_estimation(x, window_size, min_obs))
  )
  .reset_index(drop=True)
  .dropna()
)
```

```
plot_beta = (
  ggplot(beta_example,
         aes(x="month", y="beta", color="company", linetype="company")) +
  geom_line() +
  labs(x="", y="", color="", linetype="",
       title="Monthly beta estimates for example stocks") +
  scale_x_datetime(breaks=date_breaks("5 year"), labels=date_format("%Y"))
)
plot_beta.draw()
```

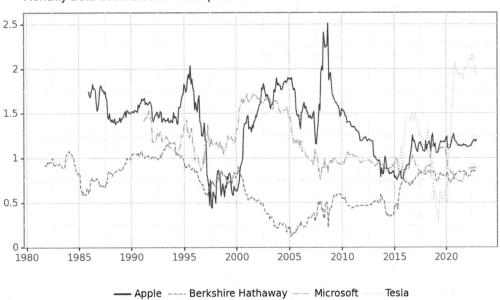

Figure 7.1: The figure shows monthly beta estimates for example stocks using five years of data. The CAPM betas are estimated with monthly data and a rolling window of length five years based on adjusted excess returns from CRSP. We use market excess returns from Kenneth French data library.

## 7.3   Parallelized Rolling-Window Estimation

Next, we perform the rolling window estimation for the entire cross-section of stocks in the CRSP sample. For that purpose, we can apply the code snippet from the example above to compute rolling window regression coefficients for all stocks. This is how to do it with the `joblib` package to use multiple cores. Note that we use `cpu_count()` to determine the number of cores available for parallelization but keep one core free for other tasks. Some machines might freeze if all cores are busy with Python jobs.

```python
def roll_capm_estimation_for_joblib(permno, group):
    """Calculate rolling CAPM estimation using joblib."""

    if "date" in group.columns:
        group = group.sort_values(by="date")
    else:
        group = group.sort_values(by="month")

    beta_values = (RollingOLS.from_formula(
        formula="ret_excess ~ mkt_excess",
        data=group,
        window=window_size,
        min_nobs=min_obs,
        missing="drop"
      )
      .fit()
      .params.get("mkt_excess")
    )

    result = pd.DataFrame(beta_values)
    result.columns = ["beta"]
    result["month"] = group["month"].values
    result["permno"] = permno

    try:
        result["date"] = group["date"].values
        result = result[
          (result.groupby("month")["date"].transform("max")) == result["date"]
        ]
    except(KeyError):
        pass

    return result

permno_groups = (returns_monthly
  .merge(valid_permnos, how="inner", on="permno")
  .groupby("permno", group_keys=False)
)

n_cores = cpu_count()-1
```

```
beta_monthly = (
  pd.concat(
    Parallel(n_jobs=n_cores)
    (delayed(roll_capm_estimation_for_joblib)(name, group)
    for name, group in permno_groups)
  )
  .dropna()
  .rename(columns={"beta": "beta_monthly"})
)
```

## 7.4   Estimating Beta Using Daily Returns

Before we provide some descriptive statistics of our beta estimates, we implement the estimation for the daily CRSP sample as well. Depending on the application, you might either use longer horizon beta estimates based on monthly data or shorter horizon estimates based on daily returns. As loading the full daily CRSP data requires relatively large amounts of memory, we split the beta estimation into smaller chunks.

First, we load the daily Fama-French market excess returns and extract the vector of dates.

```
factors_ff3_daily = pd.read_sql_query(
  sql="SELECT date, mkt_excess FROM factors_ff3_daily",
  con=tidy_finance,
  parse_dates={"date"}
)

unique_date = factors_ff3_daily["date"].unique()
```

For the daily data, we consider around three months of data (i.e., 60 trading days), require at least 50 observations, and estimate betas in batches of 500.

```
window_size = 60
min_obs = 50

permnos = list(crsp_monthly["permno"].unique().astype(str))

batch_size = 500
batches = np.ceil(len(permnos)/batch_size).astype(int)
```

We then proceed to perform the same steps as with the monthly CRSP data, just in batches: Load in daily returns, transform implicit missing returns to explicit ones, keep only valid stocks with a minimum number of rows, and parallelize the beta estimation across stocks.

```
beta_daily = []

for j in range(1, batches+1):

    permno_batch = permnos[
      ((j-1)*batch_size):(min(j*batch_size, len(permnos)))
```

```python
]

permno_batch_formatted = (
  ", ".join(f"'{permno}'" for permno in permno_batch)
)
permno_string = f"({permno_batch_formatted})"

crsp_daily_sub_query = (
  "SELECT permno, month, date, ret_excess "
    "FROM crsp_daily "
  f"WHERE permno IN {permno_string}"
)

crsp_daily_sub = pd.read_sql_query(
  sql=crsp_daily_sub_query,
  con=tidy_finance,
  dtype={"permno": int},
  parse_dates={"date", "month"}
)

valid_permnos = (crsp_daily_sub
  .groupby("permno")["permno"]
  .count()
  .reset_index(name="counts")
  .query(f"counts > {window_size}+1")
  .drop(columns="counts")
)

permno_information = (crsp_daily_sub
  .merge(valid_permnos, how="inner", on="permno")
  .groupby(["permno"])
  .aggregate(first_date=("date", "min"),
             last_date=("date", "max"),)
  .reset_index()
)

unique_permno = permno_information["permno"].unique()

all_combinations = pd.DataFrame(
  product(unique_permno, unique_date),
  columns=["permno", "date"]
)

returns_daily = (crsp_daily_sub
  .merge(all_combinations, how="right", on=["permno", "date"])
  .merge(permno_information, how="left", on="permno")
  .query("(date >= first_date) & (date <= last_date)")
  .drop(columns=["first_date", "last_date"])
  .merge(factors_ff3_daily, how="left", on="date")
)
```

```
    permno_groups = (returns_daily
      .groupby("permno", group_keys=False)
    )

    beta_daily_sub = (
      pd.concat(
        Parallel(n_jobs=n_cores)
        (delayed(roll_capm_estimation_for_joblib)(name, group)
        for name, group in permno_groups)
      )
      .dropna()
      .rename(columns={"beta": "beta_daily"})
    )

    beta_daily.append(beta_daily_sub)

    print(f"Batch {j} out of {batches} done ({(j/batches)*100:.2f}%)\n")

beta_daily = pd.concat(beta_daily)
```

---

## 7.5  Comparing Beta Estimates

What is a typical value for stock betas? To get some feeling, we illustrate the dispersion of the estimated $\hat{\beta}_i$ across different industries and across time below. Figure 7.2 shows that typical business models across industries imply different exposure to the general market economy. However, there are barely any firms that exhibit a negative exposure to the market factor.

```
beta_industries = (beta_monthly
  .merge(crsp_monthly, how="inner", on=["permno", "month"])
  .dropna(subset="beta_monthly")
  .groupby(["industry","permno"])["beta_monthly"]
  .aggregate("mean")
  .reset_index()
)

industry_order = (beta_industries
  .groupby("industry")["beta_monthly"]
  .aggregate("median")
  .sort_values()
  .index.tolist()
)

plot_beta_industries = (
  ggplot(beta_industries,
        aes(x="industry", y="beta_monthly")) +
  geom_boxplot() +
  coord_flip() +
```

```
    labs(x="", y="",
        title="Firm-specific beta distributions by industry") +
    scale_x_discrete(limits=industry_order)
)
plot_beta_industries.draw()
```

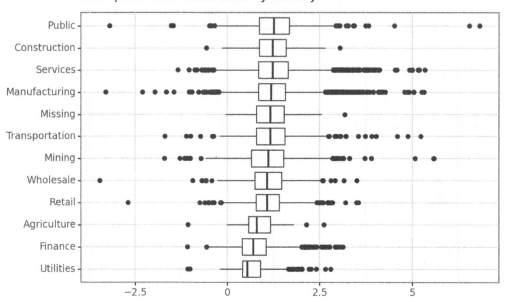

Figure 7.2: The box plots show the average firm-specific beta estimates by industry.

Next, we illustrate the time-variation in the cross-section of estimated betas. Figure 7.3 shows the monthly deciles of estimated betas (based on monthly data) and indicates an interesting pattern: First, betas seem to vary over time in the sense that during some periods, there is a clear trend across all deciles. Second, the sample exhibits periods where the dispersion across stocks increases in the sense that the lower decile decreases and the upper decile increases, which indicates that for some stocks, the correlation with the market increases, while for others it decreases. Note also here: stocks with negative betas are a rare exception.

```
beta_quantiles = (beta_monthly
  .groupby("month")["beta_monthly"]
  .quantile(q=np.arange(0.1, 1.0, 0.1))
  .reset_index()
  .rename(columns={"level_1": "quantile"})
  .assign(quantile=lambda x: (x["quantile"]*100).astype(int))
  .dropna()
)

linetypes = ["-", "--", "-.", ":"]
n_quantiles = beta_quantiles["quantile"].nunique()
```

```
plot_beta_quantiles = (
  ggplot(beta_quantiles,
         aes(x="month", y="beta_monthly",
         color="factor(quantile)", linetype="factor(quantile)")) +
  geom_line() +
  labs(x="", y="", color="", linetype="",
       title="Monthly deciles of estimated betas") +
  scale_x_datetime(breaks=date_breaks("5 year"), labels=date_format("%Y")) +
  scale_linetype_manual(
    values=[linetypes[l % len(linetypes)] for l in range(n_quantiles)]
  )
)
plot_beta_quantiles.draw()
```

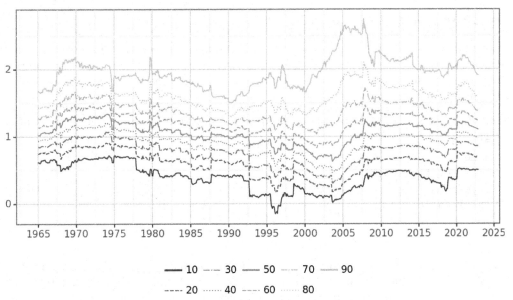

Figure 7.3: The figure shows monthly deciles of estimated betas. Each line corresponds to the monthly cross-sectional quantile of the estimated CAPM beta.

To compare the difference between daily and monthly data, we combine beta estimates to a single table. Then, we use the table to plot a comparison of beta estimates for our example stocks in Figure 7.4.

```
beta = (beta_monthly
  .get(["permno", "month", "beta_monthly"])
  .merge(beta_daily.get(["permno", "month", "beta_daily"]),
         how="outer", on=["permno", "month"])
)

beta_comparison = (beta
  .merge(examples, on="permno")
```

```
    .melt(id_vars=["permno", "month", "company"], var_name="name",
          value_vars=["beta_monthly", "beta_daily"], value_name="value")
    .dropna()
)


plot_beta_comparison = (
  ggplot(beta_comparison,
         aes(x="month", y="value", color="name")) +
  geom_line() +
  facet_wrap("~company", ncol=1) +
  labs(x="", y="", color="",
       title="Comparison of beta estimates using monthly and daily data") +
  scale_x_datetime(breaks=date_breaks("10 years"),
                   labels=date_format("%Y")) +
  theme(figure_size=(6.4, 6.4))
)
plot_beta_comparison.draw()
```

The estimates in Figure 7.4 look as expected. As you can see, the beta estimates really depend on the estimation window and data frequency. Nevertheless, one can observe a clear connection between daily and monthly betas in this example, in magnitude and the dynamics over time.

Finally, we write the estimates to our database so that we can use them in later chapters.

```
(beta.to_sql(
  name="beta",
  con=tidy_finance,
  if_exists="replace",
  index=False
  )
)
```

Whenever you perform some kind of estimation, it also makes sense to do rough plausibility tests. A possible check is to plot the share of stocks with beta estimates over time. This descriptive analysis helps us discover potential errors in our data preparation or the estimation procedure. For instance, suppose there was a gap in our output without any betas. In this case, we would have to go back and check all previous steps to find out what went wrong. Figure 7.5 does not indicate any troubles, so let us move on to the next check.

```
beta_long = (crsp_monthly
  .merge(beta, how="left", on=["permno", "month"])
  .melt(id_vars=["permno", "month"], var_name="name",
        value_vars=["beta_monthly", "beta_daily"], value_name="value")
)


beta_shares = (beta_long
  .groupby(["month", "name"])
  .aggregate(share=("value", lambda x: sum(~x.isna())/len(x)))
  .reset_index()
)
```

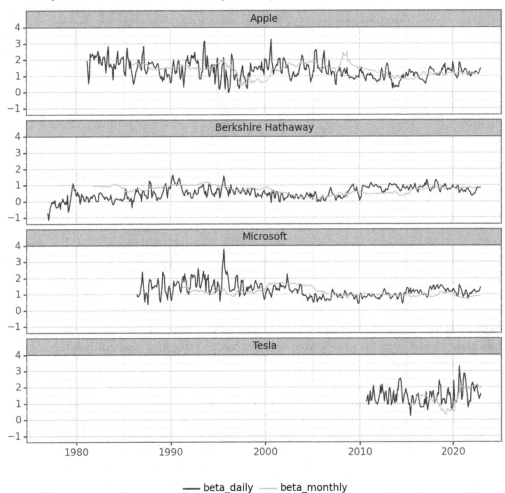

Figure 7.4: The figure shows the comparison of beta estimates using monthly and daily data. CAPM betas are computed using five years of monthly or three months of daily data. The two lines show the monthly estimates based on a rolling window for few exemplary stocks.

```
plot_beta_long = (
  ggplot(beta_shares,
         aes(x="month", y="share", color="name", linetype="name")) +
  geom_line() +
  labs(x="", y="", color="", linetype="",
       title="End-of-month share of securities with beta estimates") +
  scale_y_continuous(labels=percent_format()) +
  scale_x_datetime(breaks=date_breaks("10 year"), labels=date_format("%Y"))
)
plot_beta_long.draw()
```

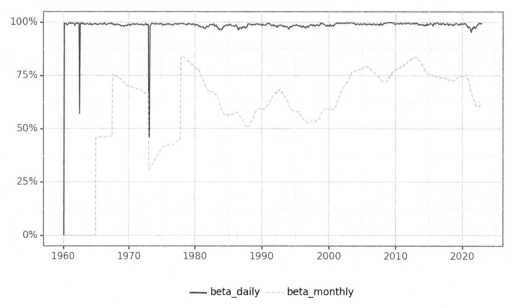

End-of-month share of securities with beta estimates

Figure 7.5: The figure shows end-of-month share of securities with beta estimates. The two lines show the share of securities with beta estimates using five years of monthly or three months of daily data.

We also encourage everyone to always look at the distributional summary statistics of variables. You can easily spot outliers or weird distributions when looking at such tables.

```
beta_long.groupby("name")["value"].describe().round(2)
```

| name | count | mean | std | min | 25% | 50% | 75% | max |
|---|---|---|---|---|---|---|---|---|
| beta_daily | 3279386.0 | 0.75 | 0.94 | -44.85 | 0.20 | 0.69 | 1.23 | 61.64 |
| beta_monthly | 2073073.0 | 1.10 | 0.70 | -8.96 | 0.64 | 1.03 | 1.47 | 10.35 |

The summary statistics also look plausible for the two estimation procedures.

Finally, since we have two different estimators for the same theoretical object, we expect the estimators to be at least positively correlated (although not perfectly as the estimators are based on different sample periods and frequencies).

```
beta.get(["beta_monthly", "beta_daily"]).corr().round(2)
```

|  | beta_monthly | beta_daily |
|---|---|---|
| beta_monthly | 1.00 | 0.32 |
| beta_daily | 0.32 | 1.00 |

Indeed, we find a positive correlation between our beta estimates. In the subsequent chapters, we mainly use the estimates based on monthly data, as most readers should be able to replicate them and should not encounter potential memory limitations that might arise with the daily data.

## 7.6 Exercises

1. Compute beta estimates based on monthly data using one, three, and five years of data and impose a minimum number of observations of 10, 28, and 48 months with return data, respectively. How strongly correlated are the estimated betas?

2. Compute beta estimates based on monthly data using five years of data and impose different numbers of minimum observations. How does the share of `permno`-`month` observations with successful beta estimates vary across the different requirements? Do you find a high correlation across the estimated betas?

3. Instead of using `joblib`, perform the beta estimation in a loop (using either monthly or daily data) for a subset of 100 permnos of your choice. Verify that you get the same results as with the parallelized code from above.

4. Filter out the stocks with negative betas. Do these stocks frequently exhibit negative betas, or do they resemble estimation errors?

5. Compute beta estimates for multi-factor models such as the Fama-French three-factor model. For that purpose, you extend your regression to

$$r_{i,t} - r_{f,t} = \alpha_i + \sum_{j=1}^{k} \beta_{i,k}(r_{j,t} - r_{f,t}) + \varepsilon_{i,t} \tag{7.2}$$

where $r_{i,t}$ are the $k$ factor returns. Thus, you estimate four parameters ($\alpha_i$ and the slope coefficients). Provide some summary statistics of the cross-section of firms and their exposure to the different factors.

# 8

## Univariate Portfolio Sorts

In this chapter, we dive into portfolio sorts, one of the most widely used statistical methodologies in empirical asset pricing (e.g., Bali et al., 2016). The key application of portfolio sorts is to examine whether one or more variables can predict future excess returns. In general, the idea is to sort individual stocks into portfolios, where the stocks within each portfolio are similar with respect to a sorting variable, such as firm size. The different portfolios then represent well-diversified investments that differ in the level of the sorting variable. You can then attribute the differences in the return distribution to the impact of the sorting variable. We start by introducing univariate portfolio sorts (which sort based on only one characteristic) and tackle bivariate sorting in Chapter 10.

A univariate portfolio sort considers only one sorting variable $x_{i,t-1}$. Here, $i$ denotes the stock and $t-1$ indicates that the characteristic is observable by investors at time $t$. The objective is to assess the cross-sectional relation between $x_{i,t-1}$ and, typically, stock excess returns $r_{i,t}$ at time $t$ as the outcome variable. To illustrate how portfolio sorts work, we use estimates for market betas from the previous chapter as our sorting variable.

The current chapter relies on the following set of Python packages.

```python
import pandas as pd
import numpy as np
import sqlite3
import statsmodels.api as sm

from plotnine import *
from mizani.formatters import percent_format
from regtabletotext import prettify_result
```

## 8.1 Data Preparation

We start with loading the required data from our SQLite database introduced in Chapter 3 and Chapter 4. In particular, we use the monthly CRSP sample as our asset universe. Once we form our portfolios, we use the Fama-French market factor returns to compute the risk-adjusted performance (i.e., alpha). `beta` is the dataframe with market betas computed in the previous chapter.

```python
tidy_finance = sqlite3.connect(database="data/tidy_finance_python.sqlite")

crsp_monthly = (pd.read_sql_query(
    sql="SELECT permno, month, ret_excess, mktcap_lag FROM crsp_monthly",
    con=tidy_finance,
```

```
    parse_dates={"month"})
)

factors_ff3_monthly = pd.read_sql_query(
  sql="SELECT month, mkt_excess FROM factors_ff3_monthly",
  con=tidy_finance,
  parse_dates={"month"}
)

beta = (pd.read_sql_query(
    sql="SELECT permno, month, beta_monthly FROM beta",
    con=tidy_finance,
    parse_dates={"month"})
)
```

## 8.2   Sorting by Market Beta

Next, we merge our sorting variable with the return data. We use the one-month
*lagged* betas as a sorting variable to ensure that the sorts rely only on information available when we create the portfolios. To lag stock beta by one month, we
add one month to the current date and join the resulting information with our return data. This procedure ensures that month $t$ information is available in month
$t + 1$. You may be tempted to simply use a call such as `crsp_monthly['beta_lag'] =
crsp_monthly.groupby('permno')['beta'].shift(1)` instead. This procedure, however,
does not work correctly if there are implicit missing values in the time series.

```
beta_lag = (beta
  .assign(month=lambda x: x["month"]+pd.DateOffset(months=1))
  .get(["permno", "month", "beta_monthly"])
  .rename(columns={"beta_monthly": "beta_lag"})
  .dropna()
)

data_for_sorts = (crsp_monthly
  .merge(beta_lag, how="inner", on=["permno", "month"])
)
```

The first step to conduct portfolio sorts is to calculate periodic breakpoints that you can use
to group the stocks into portfolios. For simplicity, we start with the median lagged market
beta as the single breakpoint. We then compute the value-weighted returns for each of the
two resulting portfolios, which means that the lagged market capitalization determines the
weight in `np.average()`.

```
beta_portfolios = (data_for_sorts
  .groupby("month")
  .apply(lambda x: (x.assign(
      portfolio=pd.qcut(
        x["beta_lag"], q=[0, 0.5, 1], labels=["low", "high"]))
```

```
      )
   )
   .reset_index(drop=True)
   .groupby(["portfolio","month"])
   .apply(lambda x: np.average(x["ret_excess"], weights=x["mktcap_lag"]))
   .reset_index(name="ret")
)
```

## 8.3 Performance Evaluation

We can construct a long-short strategy based on the two portfolios: buy the high-beta portfolio and, at the same time, short the low-beta portfolio. Thereby, the overall position in the market is net-zero, i.e., you do not need to invest money to realize this strategy in the absence of frictions.

```
beta_longshort = (beta_portfolios
   .pivot_table(index="month", columns="portfolio", values="ret")
   .reset_index()
   .assign(long_short=lambda x: x["high"]-x["low"])
)
```

We compute the average return and the corresponding standard error to test whether the long-short portfolio yields on average positive or negative excess returns. In the asset pricing literature, one typically adjusts for autocorrelation by using Newey and West (1987) $t$-statistics to test the null hypothesis that average portfolio excess returns are equal to zero. One necessary input for Newey-West standard errors is a chosen bandwidth based on the number of lags employed for the estimation. Researchers often default to choosing a pre-specified lag length of six months (which is not a data-driven approach). We do so in the fit() function by indicating the cov_type as HAC and providing the maximum lag length through an additional keywords dictionary.

```
model_fit = (sm.OLS.from_formula(
      formula="long_short ~ 1",
      data=beta_longshort
   )
   .fit(cov_type="HAC", cov_kwds={"maxlags": 6})
)
prettify_result(model_fit)
```

```
OLS Model:
long_short ~ 1

Coefficients:
            Estimate  Std. Error  t-Statistic  p-Value
Intercept       -0.0       0.001       -0.008    0.994

Summary statistics:
- Number of observations: 695
- R-squared: -0.000, Adjusted R-squared: -0.000
```

```
- F-statistic not available
```

The results indicate that we cannot reject the null hypothesis of average returns being equal to zero. Our portfolio strategy using the median as a breakpoint does not yield any abnormal returns. Is this finding surprising if you reconsider the CAPM? It certainly is. The CAPM yields that the high-beta stocks should yield higher expected returns. Our portfolio sort implicitly mimics an investment strategy that finances high-beta stocks by shorting low-beta stocks. Therefore, one should expect that the average excess returns yield a return that is above the risk-free rate.

## 8.4    Functional Programming for Portfolio Sorts

Now, we take portfolio sorts to the next level. We want to be able to sort stocks into an arbitrary number of portfolios. For this case, functional programming is very handy: we define a function that gives us flexibility concerning which variable to use for the sorting, denoted by `sorting_variable`. We use `np.quantile()` to compute breakpoints for `n_portfolios`. Then, we assign portfolios to stocks using the `pd.cut()` function. The output of the following function is a new column that contains the number of the portfolio to which a stock belongs.

In some applications, the variable used for the sorting might be clustered (e.g., at a lower bound of 0). Then, multiple breakpoints may be identical, leading to empty portfolios. Similarly, some portfolios might have a very small number of stocks at the beginning of the sample. Cases where the number of portfolio constituents differs substantially due to the distribution of the characteristics require careful consideration and, depending on the application, might require customized sorting approaches.

```python
def assign_portfolio(data, sorting_variable, n_portfolios):
    """Assign portfolios to a bin between breakpoints."""

    breakpoints = np.quantile(
      data[sorting_variable].dropna(),
      np.linspace(0, 1, n_portfolios + 1),
      method="linear"
    )

    assigned_portfolios = pd.cut(
      data[sorting_variable],
      bins=breakpoints,
      labels=range(1, breakpoints.size),
      include_lowest=True,
      right=False
    )

    return assigned_portfolios
```

We can use the above function to sort stocks into ten portfolios each month using lagged betas and compute value-weighted returns for each portfolio. Note that we transform the portfolio column to a factor variable because it provides more convenience for the figure construction below.

```python
beta_portfolios = (data_for_sorts
  .groupby("month")
  .apply(lambda x: x.assign(
      portfolio=assign_portfolio(x, "beta_lag", 10)
    )
  )
  .reset_index(drop=True)
  .groupby(["portfolio", "month"])
  .apply(lambda x: x.assign(
      ret=np.average(x["ret_excess"], weights=x["mktcap_lag"])
    )
  )
  .reset_index(drop=True)
  .merge(factors_ff3_monthly, how="left", on="month")
)
```

## 8.5   More Performance Evaluation

In the next step, we compute summary statistics for each beta portfolio. Namely, we compute CAPM-adjusted alphas, the beta of each beta portfolio, and average returns.

```python
beta_portfolios_summary = (beta_portfolios
  .groupby("portfolio")
  .apply(lambda x: x.assign(
      alpha=sm.OLS.from_formula(
          formula="ret ~ 1 + mkt_excess",
          data=x
        ).fit().params[0],
      beta=sm.OLS.from_formula(
          formula="ret ~ 1 + mkt_excess",
          data=x
        ).fit().params[1],
    ret=x["ret"].mean()
  ).tail(1)
  )
  .reset_index(drop=True)
  .get(["portfolio", "alpha", "beta", "ret"])
)
```

Figure 8.1 illustrates the CAPM alphas of beta-sorted portfolios. It shows that low-beta portfolios tend to exhibit positive alphas, while high-beta portfolios exhibit negative alphas.

```python
plot_beta_portfolios_summary = (
  ggplot(beta_portfolios_summary,
         aes(x="portfolio", y="alpha", fill="portfolio")) +
  geom_bar(stat="identity") +
```

```
    labs(x="Portfolio", y="CAPM alpha", fill="Portfolio",
         title="CAPM alphas of beta-sorted portfolios") +
    scale_y_continuous(labels=percent_format()) +
    theme(legend_position="none")
)
plot_beta_portfolios_summary.draw()
```

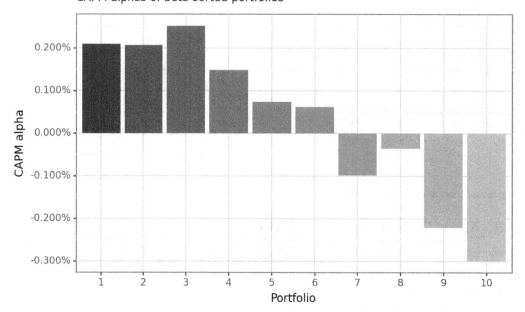

Figure 8.1: The figure shows CAPM alphas of beta-sorted portfolios. Portfolios are sorted into deciles each month based on their estimated CAPM beta. The bar charts indicate the CAPM alpha of the resulting portfolio returns during the entire CRSP period.

These results suggest a negative relation between beta and future stock returns, which contradicts the predictions of the CAPM. According to the CAPM, returns should increase with beta across the portfolios and risk-adjusted returns should be statistically indistinguishable from zero.

## 8.6   Security Market Line and Beta Portfolios

The CAPM predicts that our portfolios should lie on the security market line (SML). The slope of the SML is equal to the market risk premium and reflects the risk-return trade-off at any given time. Figure 8.2 illustrates the security market line: We see that (not surprisingly) the high-beta portfolio returns have a high correlation with the market returns. However, it seems like the average excess returns for high-beta stocks are lower than what the security market line implies would be an "appropriate" compensation for the high market risk.

```
sml_capm = (sm.OLS.from_formula(
    formula="ret ~ 1 + beta",
    data=beta_portfolios_summary
  )
  .fit()
  .params
)


plot_sml_capm = (
  ggplot(beta_portfolios_summary,
         aes(x="beta", y="ret", color="portfolio")) +
  geom_point() +
  geom_abline(intercept=0,
              slope=factors_ff3_monthly["mkt_excess"].mean(),
              linetype="solid") +
  geom_abline(intercept=sml_capm["Intercept"],
              slope=sml_capm["beta"],
              linetype="dashed") +
  labs(x="Beta", y="Excess return", color="Portfolio",
       title="Average portfolio excess returns and beta estimates") +
  scale_x_continuous(limits=(0, 2)) +
  scale_y_continuous(labels=percent_format(),
                     limits=(0, factors_ff3_monthly["mkt_excess"].mean()*2))
)
plot_sml_capm.draw()
```

To provide more evidence against the CAPM predictions, we again form a long-short strategy that buys the high-beta portfolio and shorts the low-beta portfolio.

```
beta_longshort = (beta_portfolios
  .assign(
    portfolio=lambda x: (
      x["portfolio"].apply(
        lambda y: "high" if y == x["portfolio"].max()
        else ("low" if y == x["portfolio"].min()
        else y)
      )
    )
  )
  .query("portfolio in ['low', 'high']")
  .pivot_table(index="month", columns="portfolio", values="ret")
  .assign(long_short=lambda x: x["high"]-x["low"])
  .merge(factors_ff3_monthly, how="left", on="month")
)
```

Again, the resulting long-short strategy does not exhibit statistically significant returns.

```
model_fit = (sm.OLS.from_formula(
    formula="long_short ~ 1",
    data=beta_longshort
  )
  .fit(cov_type="HAC", cov_kwds={"maxlags": 1})
```

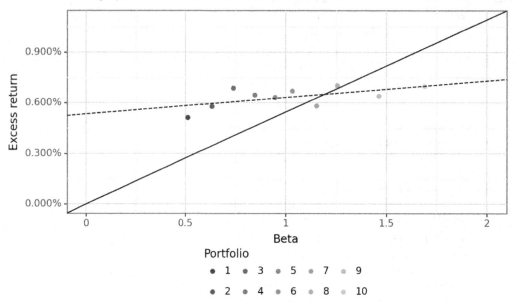

Figure 8.2: The figure shows average portfolio excess returns and beta estimates. Excess returns are computed as CAPM alphas of the beta-sorted portfolios. The horizontal axis indicates the CAPM beta of the resulting beta-sorted portfolio return time series. The dashed line indicates the slope coefficient of a linear regression of excess returns on portfolio betas.

```
)
prettify_result(model_fit)

OLS Model:
long_short ~ 1

Coefficients:
          Estimate  Std. Error  t-Statistic  p-Value
Intercept    0.002       0.003        0.549    0.583

Summary statistics:
- Number of observations: 695
- R-squared: 0.000, Adjusted R-squared: 0.000
- F-statistic not available
```

However, controlling for the effect of beta, the long-short portfolio yields a statistically significant negative CAPM-adjusted alpha, although the average excess stock returns should be zero according to the CAPM. The results thus provide no evidence in support of the CAPM. The negative value has been documented as the so-called betting-against-beta factor (Frazzini and Pedersen, 2014). Betting-against-beta corresponds to a strategy that shorts high-beta stocks and takes a (levered) long position in low-beta stocks. If borrowing constraints prevent investors from taking positions on the security market line they are instead incentivized to buy high-beta stocks, which leads to a relatively higher price (and therefore lower expected returns than implied by the CAPM) for such high-beta stocks. As a result, the betting-against-beta strategy earns from providing liquidity to capital-constrained investors with lower risk aversion.

```
model_fit = (sm.OLS.from_formula(
    formula="long_short ~ 1 + mkt_excess",
    data=beta_longshort
  )
  .fit(cov_type="HAC", cov_kwds={"maxlags": 1})
)
prettify_result(model_fit)
```

```
OLS Model:
long_short ~ 1 + mkt_excess
```

```
Coefficients:
            Estimate  Std. Error  t-Statistic  p-Value
Intercept    -0.004      0.002       -1.931     0.053
mkt_excess    1.129      0.069       16.392     0.000
```

```
Summary statistics:
- Number of observations: 695
- R-squared: 0.439, Adjusted R-squared: 0.438
- F-statistic: 268.700 on 1 and 693 DF, p-value: 0.000
```

Figure 8.3 shows the annual returns of the extreme beta portfolios we are mainly interested in. The figure illustrates no consistent striking patterns over the last years; each portfolio exhibits periods with positive and negative annual returns.

```
beta_longshort_year = (beta_longshort
  .assign(year=lambda x: x["month"].dt.year)
  .groupby("year")
  .aggregate(
    low=("low", lambda x: 1-(1+x).prod()),
    high=("high", lambda x: 1-(1+x).prod()),
    long_short=("long_short", lambda x: 1-(1+x).prod())
  )
  .reset_index()
  .melt(id_vars="year", var_name="name", value_name="value")
)

plot_beta_longshort_year = (
  ggplot(beta_longshort_year,
        aes(x="year", y="value", fill="name")) +
  geom_col(position='dodge') +
  facet_wrap("~name", ncol=1) +
  labs(x="", y="", title="Annual returns of beta portfolios") +
  scale_color_discrete(guide=False) +
  scale_y_continuous(labels=percent_format()) +
  theme(legend_position="none")
)
plot_beta_longshort_year.draw()
```

Annual returns of beta portfolios

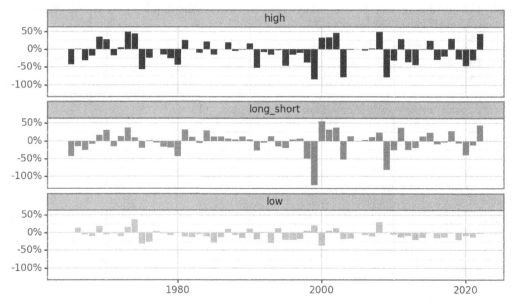

Figure 8.3: The figure shows annual returns of beta portfolios. We construct portfolios by sorting stocks into high and low based on their estimated CAPM beta. Long short indicates a strategy that goes long into high beta stocks and short low beta stocks.

Overall, this chapter shows how functional programming can be leveraged to form an arbitrary number of portfolios using any sorting variable and how to evaluate the performance of the resulting portfolios. In the next chapter, we dive deeper into the many degrees of freedom that arise in the context of portfolio analysis.

## 8.7   Exercises

1. Take the two long-short beta strategies based on different numbers of portfolios and compare the returns. Is there a significant difference in returns? How do the Sharpe ratios compare between the strategies? Find one additional portfolio evaluation statistic and compute it.
2. We plotted the alphas of the ten beta portfolios above. Write a function that tests these estimates for significance. Which portfolios have significant alphas?
3. The analysis here is based on betas from monthly returns. However, we also computed betas from daily returns. Re-run the analysis and point out differences in the results.
4. Given the results in this chapter, can you define a long-short strategy that yields positive abnormal returns (i.e., alphas)? Plot the cumulative excess return of your strategy and the market excess return for comparison.

# 9

## _Size Sorts and p-Hacking_

In this chapter, we continue with portfolio sorts in a univariate setting. Yet, we consider firm size as a sorting variable, which gives rise to a well-known return factor: the size premium. The size premium arises from buying small stocks and selling large stocks. Prominently, Fama and French (1993) include it as a factor in their three-factor model. Apart from that, asset managers commonly include size as a key firm characteristic when making investment decisions.

We also introduce new choices in the formation of portfolios. In particular, we discuss listing exchanges, industries, weighting regimes, and periods. These choices matter for the portfolio returns and result in different size premiums (see Hasler, 2021, Soebhag et al. (2022), and Walter et al. (2022) for more insights into decision nodes and their effect on premiums). Exploiting these ideas to generate favorable results is called p-hacking. There is arguably a thin line between p-hacking and conducting robustness tests. Our purpose here is to illustrate the substantial variation that can arise along the evidence-generating process.

The chapter relies on the following set of Python packages:

```python
import pandas as pd
import numpy as np
import sqlite3

from plotnine import *
from mizani.formatters import percent_format
from itertools import product
from joblib import Parallel, delayed, cpu_count
```

Compared to previous chapters, we introduce `itertools`, which is a component of the Python standard library and provides fast, memory-efficient tools for working with iterators.

## 9.1 Data Preparation

First, we retrieve the relevant data from our SQLite database introduced in Chapter 3 and Chapter 4. Firm size is defined as market equity in most asset pricing applications that we retrieve from CRSP. We further use the Fama-French factor returns for performance evaluation.

```python
tidy_finance = sqlite3.connect(
  database="data/tidy_finance_python.sqlite"
)
```

```
crsp_monthly = pd.read_sql_query(
  sql="SELECT * FROM crsp_monthly",
  con=tidy_finance,
  parse_dates={"month"}
)

factors_ff3_monthly = pd.read_sql_query(
  sql="SELECT * FROM factors_ff3_monthly",
  con=tidy_finance,
  parse_dates={"month"}
)
```

## 9.2   Size Distribution

Before we build our size portfolios, we investigate the distribution of the variable *firm size*. Visualizing the data is a valuable starting point to understand the input to the analysis. Figure 9.1 shows the fraction of total market capitalization concentrated in the largest firm. To produce this graph, we create monthly indicators that track whether a stock belongs to the largest x percent of the firms. Then, we aggregate the firms within each bucket and compute the buckets' share of total market capitalization.

Figure 9.1 shows that the largest 1 percent of firms cover up to 50 percent of the total market capitalization, and holding just the 25 percent largest firms in the CRSP universe essentially replicates the market portfolio. The distribution of firm size thus implies that the largest firms of the market dominate many small firms whenever we use value-weighted benchmarks.

```
market_cap_concentration = (crsp_monthly
  .groupby("month")
  .apply(lambda x: x.assign(
    top01=(x["mktcap"] >= np.quantile(x["mktcap"], 0.99)),
    top05=(x["mktcap"] >= np.quantile(x["mktcap"], 0.95)),
    top10=(x["mktcap"] >= np.quantile(x["mktcap"], 0.90)),
    top25=(x["mktcap"] >= np.quantile(x["mktcap"], 0.75)))
  )
  .reset_index(drop=True)
  .groupby("month")
  .apply(lambda x: pd.Series({
    "Largest 1%": x["mktcap"][x["top01"]].sum()/x["mktcap"].sum(),
    "Largest 5%": x["mktcap"][x["top05"]].sum()/x["mktcap"].sum(),
    "Largest 10%": x["mktcap"][x["top10"]].sum()/x["mktcap"].sum(),
    "Largest 25%": x["mktcap"][x["top25"]].sum()/x["mktcap"].sum()
    })
  )
  .reset_index()
  .melt(id_vars="month", var_name="name", value_name="value")
)
```

```
plot_market_cap_concentration = (
  ggplot(market_cap_concentration,
         aes(x="month", y="value",
         color="name", linetype="name")) +
  geom_line() +
  scale_y_continuous(labels=percent_format()) +
  scale_x_date(name="", date_labels="%Y") +
  labs(x="", y="", color="", linetype="",
       title=("Percentage of total market capitalization in "
              "largest stocks")) +
  theme(legend_title=element_blank())
)
plot_market_cap_concentration.draw()
```

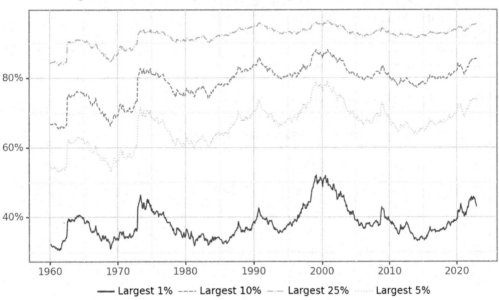

Figure 9.1: The figure shows the percentage of total market capitalization in largest stocks. We report the aggregate market capitalization of all stocks that belong to the 1, 5, 10, and 25 percent quantile of the largest firms in the monthly cross-section relative to the market capitalization of all stocks during the month.

Next, firm sizes also differ across listing exchanges. The primary listings of stocks were important in the past and are potentially still relevant today. Figure 9.2 shows that the New York Stock Exchange (NYSE) was and still is the largest listing exchange in terms of market capitalization. More recently, NASDAQ has gained relevance as a listing exchange. Do you know what the small peak in NASDAQ's market cap around the year 2000 was?

```
market_cap_share = (crsp_monthly
  .groupby(["month", "exchange"])
  .aggregate({"mktcap": "sum"})
  .reset_index(drop=False)
```

```
    .groupby("month")
    .apply(lambda x:
      x.assign(total_market_cap=lambda x: x["mktcap"].sum(),
               share=lambda x: x["mktcap"]/x["total_market_cap"]
               )
     )
    .reset_index(drop=True)
)

plot_market_cap_share = (
  ggplot(market_cap_share,
         aes(x="month", y="share",
             fill="exchange", color="exchange")) +
  geom_area(position="stack", stat="identity", alpha=0.5) +
  geom_line(position="stack") +
  scale_y_continuous(labels=percent_format()) +
  scale_x_date(name="", date_labels="%Y") +
  labs(x="", y="", fill="", color="",
       title="Share of total market capitalization per listing exchange") +
  theme(legend_title=element_blank())
)
plot_market_cap_share.draw()
```

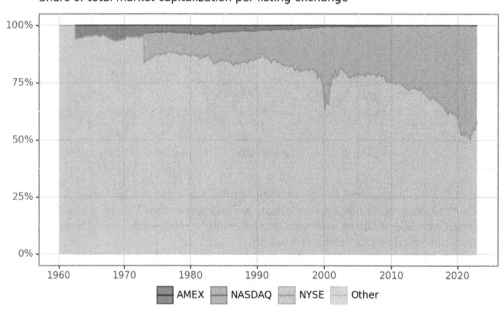

Figure 9.2: The figure shows the share of total market capitalization per listing exchange. Years are on the horizontal axis and the corresponding share of total market capitalization per listing exchange on the vertical axis.

Finally, we consider the distribution of firm size across listing exchanges and create summary statistics. The function `describe()` does not include all statistics we are interested in, which is why we create the function `compute_summary()` that adds the standard deviation and the number of observations. Then, we apply it to the most current month of our CRSP data on each listing exchange. We also add a row with the overall summary statistics.

The resulting table shows that firms listed on NYSE in December 2021 are significantly larger on average than firms listed on the other exchanges. Moreover, NASDAQ lists the largest number of firms. This discrepancy between firm sizes across listing exchanges motivated researchers to form breakpoints exclusively on the NYSE sample and apply those breakpoints to all stocks. In the following, we use this distinction to update our portfolio sort procedure.

```python
def compute_summary(data, variable, filter_variable, percentiles):
    """Compute summary statistics for a given variable and percentiles."""

    summary = (data
      .get([filter_variable, variable])
      .groupby(filter_variable)
      .describe(percentiles=percentiles)
    )

    summary.columns = summary.columns.droplevel(0)

    summary_overall = (data
      .get(variable)
      .describe(percentiles=percentiles)
    )

    summary.loc["Overall", :] = summary_overall

    return summary.round(0)

compute_summary(
  crsp_monthly[crsp_monthly["month"] == crsp_monthly["month"].max()],
  variable="mktcap",
  filter_variable="exchange",
  percentiles=[0.05, 0.5, 0.95]
)
```

| exchange | count | mean | std | min | 5% | 50% | 95% | max |
|---|---|---|---|---|---|---|---|---|
| AMEX | 162.0 | 407.0 | 2954.0 | 3.0 | 5.0 | 45.0 | 947.0 | 37289.0 |
| NASDAQ | 2777.0 | 5684.0 | 58811.0 | 1.0 | 8.0 | 265.0 | 13040.0 | 2058404.0 |
| NYSE | 1355.0 | 16299.0 | 47361.0 | 11.0 | 112.0 | 2818.0 | 67924.0 | 495373.0 |
| Other | 1.0 | 13310.0 | NaN | 13310.0 | 13310.0 | 13310.0 | 13310.0 | 13310.0 |
| Overall | 4295.0 | 8836.0 | 54500.0 | 1.0 | 11.0 | 487.0 | 34170.0 | 2058404.0 |

## 9.3    Univariate Size Portfolios with Flexible Breakpoints

In Chapter 8, we construct portfolios with a varying number of breakpoints and different sorting variables. Here, we extend the framework such that we compute breakpoints on a subset of the data, for instance, based on selected listing exchanges. In published asset pricing articles, many scholars compute sorting breakpoints only on NYSE-listed stocks. These NYSE-specific breakpoints are then applied to the entire universe of stocks.

To replicate the NYSE-centered sorting procedure, we introduce `exchanges` as an argument in our `assign_portfolio()` function from Chapter 8. The exchange-specific argument then enters in the filter `data["exchanges"].isin(exchanges)`. For example, if `exchanges='NYSE'` is specified, only stocks listed on NYSE are used to compute the breakpoints. Alternatively, you could specify `exchanges=["NYSE", "NASDAQ", "AMEX"]`, which keeps all stocks listed on either of these exchanges.

```python
def assign_portfolio(data, exchanges, sorting_variable, n_portfolios):
    """Assign portfolio for a given sorting variable."""

    breakpoints = (data
      .query(f"exchange in {exchanges}")
      .get(sorting_variable)
      .quantile(np.linspace(0, 1, num=n_portfolios+1),
               interpolation="linear")
      .drop_duplicates()
    )
    breakpoints.iloc[[0, -1]] = [-np.Inf, np.Inf]

    assigned_portfolios = pd.cut(
      data[sorting_variable],
      bins=breakpoints,
      labels=range(1, breakpoints.size),
      include_lowest=True,
      right=False
    )

    return assigned_portfolios
```

## 9.4    Weighting Schemes for Portfolios

Apart from computing breakpoints on different samples, researchers often use different portfolio weighting schemes. So far, we weighted each portfolio constituent by its relative market equity of the previous period. This protocol is called *value-weighting*. The alternative protocol is *equal-weighting*, which assigns each stock's return the same weight, i.e., a simple average of the constituents' returns. Notice that equal-weighting is difficult in practice as the portfolio manager needs to rebalance the portfolio monthly, while value-weighting is a truly passive investment.

We implement the two weighting schemes in the function `compute_portfolio_returns()` that takes a logical argument to weight the returns by firm value. Additionally, the long-short portfolio is long in the smallest firms and short in the largest firms, consistent with research showing that small firms outperform their larger counterparts. Apart from these two changes, the function is similar to the procedure in Chapter 8.

```python
def calculate_returns(data, value_weighted):
    """Calculate (value-weighted) returns."""

    if value_weighted:
        return np.average(data["ret_excess"], weights=data["mktcap_lag"])
    else:
        return data["ret_excess"].mean()


def compute_portfolio_returns(n_portfolios=10,
                              exchanges=["NYSE", "NASDAQ", "AMEX"],
                              value_weighted=True,
                              data=crsp_monthly):
    """Compute (value-weighted) portfolio returns."""

    returns = (data
      .groupby("month")
      .apply(lambda x: x.assign(
        portfolio=assign_portfolio(x, exchanges,
                                   "mktcap_lag", n_portfolios))
      )
      .reset_index(drop=True)
      .groupby(["portfolio", "month"])
      .apply(lambda x: x.assign(
        ret=calculate_returns(x, value_weighted))
      )
      .reset_index(drop=True)
      .groupby("month")
      .apply(lambda x:
        pd.Series({"size_premium": x.loc[x["portfolio"].idxmin(), "ret"]-
          x.loc[x["portfolio"].idxmax(), "ret"]}))
      .reset_index(drop=True)
      .aggregate({"size_premium": "mean"})
    )

    return returns
```

To see how the function `compute_portfolio_returns()` works, we consider a simple median breakpoint example with value-weighted returns. We are interested in the effect of restricting listing exchanges on the estimation of the size premium. In the first function call, we compute returns based on breakpoints from all listing exchanges. Then, we computed returns based on breakpoints from NYSE-listed stocks.

```python
ret_all = compute_portfolio_returns(
  n_portfolios=2,
  exchanges=["NYSE", "NASDAQ", "AMEX"],
  value_weighted=True,
```

```
  data=crsp_monthly
)

ret_nyse = compute_portfolio_returns(
  n_portfolios=2,
  exchanges=["NYSE"],
  value_weighted=True,
  data=crsp_monthly
)


data = pd.DataFrame([ret_all*100, ret_nyse*100],
                index=["NYSE, NASDAQ & AMEX", "NYSE"])
data.columns = ["Premium"]
data.round(2)
```

|                     | Premium |
|---------------------|---------|
| NYSE, NASDAQ & AMEX | 0.08    |
| NYSE                | 0.16    |

The table shows that the size premium is more than 60 percent larger if we consider only stocks from NYSE to form the breakpoint each month. The NYSE-specific breakpoints are larger, and there are more than 50 percent of the stocks in the entire universe in the resulting small portfolio because NYSE firms are larger on average. The impact of this choice is not negligible.

## 9.5   P-Hacking and Non-Standard Errors

Since the choice of the listing exchange has a significant impact, the next step is to investigate the effect of other data processing decisions researchers have to make along the way. In particular, any portfolio sort analysis has to decide at least on the number of portfolios, the listing exchanges to form breakpoints, and equal- or value-weighting. Further, one may exclude firms that are active in the finance industry or restrict the analysis to some parts of the time series. All of the variations of these choices that we discuss here are part of scholarly articles published in the top finance journals. We refer to Walter et al. (2022) for an extensive set of other decision nodes at the discretion of researchers.

The intention of this application is to show that the different ways to form portfolios result in different estimated size premiums. Despite the effects of this multitude of choices, there is no correct way. It should also be noted that none of the procedures is wrong. The aim is simply to illustrate the changes that can arise due to the variation in the evidence-generating process (Menkveld et al., 2021). The term *non-standard errors* refers to the variation due to (suitable) choices made by researchers. Interestingly, in a large-scale study, Menkveld et al. (2021) find that the magnitude of non-standard errors are similar to the estimation uncertainty based on a chosen model. This shows how important it is to adjust for the seemingly innocent choices in the data preparation and evaluation workflow. Moreover, it seems that this methodology-related uncertainty should be embraced rather than hidden away.

From a malicious perspective, these modeling choices give the researcher multiple *chances* to find statistically significant results. Yet this is considered *p-hacking*, which renders the statistical inference invalid due to multiple testing (Harvey et al., 2016).

Nevertheless, the multitude of options creates a problem since there is no single correct way of sorting portfolios. How should a researcher convince a reader that their results do not come from a p-hacking exercise? To circumvent this dilemma, academics are encouraged to present evidence from different sorting schemes as *robustness tests* and report multiple approaches to show that a result does not depend on a single choice. Thus, the robustness of premiums is a key feature.

Below, we conduct a series of robustness tests, which could also be interpreted as a p-hacking exercise. To do so, we examine the size premium in different specifications presented in the table `p_hacking_setup`. The function `itertools.product()` produces all possible permutations of its arguments. Note that we use the argument `data` to exclude financial firms and truncate the time series.

```
n_portfolios = [2, 5, 10]
exchanges = [["NYSE"], ["NYSE", "NASDAQ", "AMEX"]]
value_weighted = [True, False]
data = [
  crsp_monthly,
  crsp_monthly[crsp_monthly["industry"] != "Finance"],
  crsp_monthly[crsp_monthly["month"] < "1990-06-01"],
  crsp_monthly[crsp_monthly["month"] >= "1990-06-01"],
]
p_hacking_setup = list(
  product(n_portfolios, exchanges, value_weighted, data)
)
```

To speed the computation up, we parallelize the (many) different sorting procedures, as in Chapter 7 using the `joblib` package. Finally, we report the resulting size premiums in descending order. There are indeed substantial size premiums possible in our data, in particular when we use equal-weighted portfolios.

```
n_cores = cpu_count()-1
p_hacking_results = pd.concat(
  Parallel(n_jobs=n_cores)
  (delayed(compute_portfolio_returns)(x, y, z, w)
    for x, y, z, w in p_hacking_setup)
)
p_hacking_results = p_hacking_results.reset_index(name="size_premium")
```

## 9.6 Size-Premium Variation

We provide a graph in Figure 9.3 that shows the different premiums. The figure also shows the relation to the average Fama-French SMB (small minus big) premium used in the literature, which we include as a dotted vertical line.

```
p_hacking_results_figure = (
  ggplot(p_hacking_results,
         aes(x="size_premium")) +
  geom_histogram(bins=len(p_hacking_results)) +
  scale_x_continuous(labels=percent_format()) +
  labs(x="", y="",
       title="Distribution of size premiums for various sorting choices") +
  geom_vline(aes(xintercept=factors_ff3_monthly["smb"].mean()),
             linetype="dashed")
)
p_hacking_results_figure.draw()
```

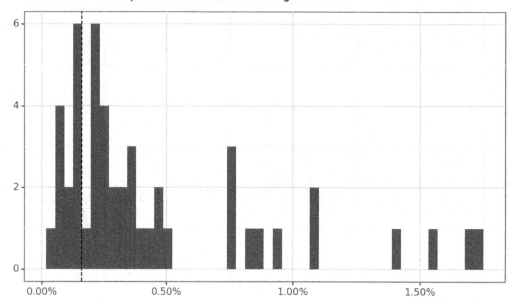

Figure 9.3: The figure shows the distribution of size premiums for various sorting choices. The dashed vertical line indicates the average Fama-French SMB premium.

## 9.7  Exercises

1. We gained several insights on the size distribution above. However, we did not analyze the average size across listing exchanges and industries. Which listing exchanges/industries have the largest firms? Plot the average firm size for the three listing exchanges over time. What do you conclude?

2. We compute breakpoints but do not take a look at them in the exposition above. This might cover potential data errors. Plot the breakpoints for ten size portfolios over time. Then, take the difference between the two extreme portfolios and plot it. Describe your results.

3. The returns that we analyze above do not account for differences in the exposure to market risk, i.e., the CAPM beta. Change the function `compute_portfolio_returns()` to output the CAPM alpha or beta instead of the average excess return.
4. While you saw the spread in returns from the p-hacking exercise, we did not show which choices led to the largest effects. Find a way to investigate which choice variable has the largest impact on the estimated size premium.
5. We computed several size premiums, but they do not follow the definition of Fama and French (1993). Which of our approaches comes closest to their SMB premium?

# 10

# *Value and Bivariate Sorts*

In this chapter, we extend the univariate portfolio analysis of Chapter 8 to bivariate sorts, which means we assign stocks to portfolios based on two characteristics. Bivariate sorts are regularly used in the academic asset pricing literature and are the basis for factors in the Fama-French three-factor model. However, some scholars also use sorts with three grouping variables. Conceptually, portfolio sorts are easily applicable in higher dimensions.

We form portfolios on firm size and the book-to-market ratio. To calculate book-to-market ratios, accounting data is required, which necessitates additional steps during portfolio formation. In the end, we demonstrate how to form portfolios on two sorting variables using so-called independent and dependent portfolio sorts.

The current chapter relies on this set of Python packages.

```python
import pandas as pd
import numpy as np
import datetime as dt
import sqlite3
```

Compared to previous chapters, we introduce the `datetime` module that is part of the Python standard library for manipulating dates.

## 10.1  Data Preparation

First, we load the necessary data from our SQLite database introduced in Chapter 3 and Chapter 4. We conduct portfolio sorts based on the CRSP sample but keep only the necessary columns in our memory. We use the same data sources for firm size as in Chapter 9.

```python
tidy_finance = sqlite3.connect(database="data/tidy_finance_python.sqlite")

crsp_monthly = (pd.read_sql_query(
    sql=("SELECT permno, gvkey, month, ret_excess, mktcap, "
         "mktcap_lag, exchange FROM crsp_monthly"),
    con=tidy_finance,
    parse_dates={"month"})
  .dropna()
)
```

Further, we utilize accounting data. The most common source of accounting data is Compustat. We only need book equity data in this application, which we select from our database.

Additionally, we convert the variable `datadate` to its monthly value, as we only consider monthly returns here and do not need to account for the exact date.

```
book_equity = (pd.read_sql_query(
    sql="SELECT gvkey, datadate, be FROM compustat",
    con=tidy_finance,
    parse_dates={"datadate"})
  .dropna()
  .assign(
    month=lambda x: (
      pd.to_datetime(x["datadate"]).dt.to_period("M").dt.to_timestamp()
    )
  )
)
```

## 10.2  Book-to-Market Ratio

A fundamental problem in handling accounting data is the *look-ahead bias*; we must not include data in forming a portfolio that was not available knowledge at the time. Of course, researchers have more information when looking into the past than agents actually had at that moment. However, abnormal excess returns from a trading strategy should not rely on an information advantage because the differential cannot be the result of informed agents' trades. Hence, we have to lag accounting information.

We continue to lag market capitalization and firm size by one month. Then, we compute the book-to-market ratio, which relates a firm's book equity to its market equity. Firms with high (low) book-to-market ratio are called value (growth) firms. After matching the accounting and market equity information from the same month, we lag book-to-market by six months. This is a sufficiently conservative approach because accounting information is usually released well before six months pass. However, in the asset pricing literature, even longer lags are used as well.[1]

Having both variables, i.e., firm size lagged by one month and book-to-market lagged by six months, we merge these sorting variables to our returns using the `sorting_date`-column created for this purpose. The final step in our data preparation deals with differences in the frequency of our variables. Returns and firm size are recorded monthly. Yet, the accounting information is only released on an annual basis. Hence, we only match book-to-market to one month per year and have eleven empty observations. To solve this frequency issue, we carry the latest book-to-market ratio of each firm to the subsequent months, i.e., we fill the missing observations with the most current report. This is done via the `fillna(method="ffill")`-function after sorting by date and firm (which we identify by `permno` and `gvkey`) and on a firm basis (which we do by `.groupby()` as usual). We filter out all observations with accounting data that is older than a year. As the last step, we remove all rows with missing entries because the returns cannot be matched to any annual report.

```
me = (crsp_monthly
  .assign(sorting_date=lambda x: x["month"]+pd.DateOffset(months=1))
```

---

[1]The definition of a time lag is another choice a researcher has to make, similar to breakpoint choices as we describe in Chapter 9.

```
  .rename(columns={"mktcap": "me"})
  .get(["permno", "sorting_date", "me"])
)

bm = (book_equity
  .merge(crsp_monthly, how="inner", on=["gvkey", "month"])
  .assign(bm=lambda x: x["be"]/x["mktcap"],
          sorting_date=lambda x: x["month"]+pd.DateOffset(months=6))
  .assign(comp_date=lambda x: x["sorting_date"])
  .get(["permno", "gvkey", "sorting_date", "comp_date", "bm"])
)

data_for_sorts = (crsp_monthly
  .merge(bm,
         how="left",
         left_on=["permno", "gvkey", "month"],
         right_on=["permno", "gvkey", "sorting_date"])
  .merge(me,
         how="left",
         left_on=["permno", "month"],
         right_on=["permno", "sorting_date"])
  .get(["permno", "gvkey", "month", "ret_excess",
        "mktcap_lag", "me", "bm", "exchange", "comp_date"])
)

data_for_sorts = (data_for_sorts
  .sort_values(by=["permno", "gvkey", "month"])
  .groupby(["permno", "gvkey"])
  .apply(lambda x: x.assign(
     bm=x["bm"].fillna(method="ffill"),
     comp_date=x["comp_date"].fillna(method="ffill")
   )
  )
  .reset_index(drop=True)
  .assign(threshold_date = lambda x: (x["month"]-pd.DateOffset(months=12)))
  .query("comp_date > threshold_date")
  .drop(columns=["comp_date", "threshold_date"])
  .dropna()
)
```

The last step of preparation for the portfolio sorts is the computation of breakpoints. We continue to use the same function, allowing for the specification of exchanges to be used for the breakpoints. Additionally, we reintroduce the argument `sorting_variable` into the function for defining different sorting variables.

```
def assign_portfolio(data, exchanges, sorting_variable, n_portfolios):
    """Assign portfolio for a given sorting variable."""

    breakpoints = (data
      .query(f"exchange in {exchanges}")
      .get(sorting_variable)
```

```
    .quantile(np.linspace(0, 1, num=n_portfolios+1),
             interpolation="linear")
    .drop_duplicates()
)
breakpoints.iloc[0] = -np.Inf
breakpoints.iloc[breakpoints.size-1] = np.Inf

assigned_portfolios = pd.cut(
  data[sorting_variable],
  bins=breakpoints,
  labels=range(1, breakpoints.size),
  include_lowest=True,
  right=False
)

return assigned_portfolios
```

After these data preparation steps, we present bivariate portfolio sorts on an independent and dependent basis.

## 10.3   Independent Sorts

Bivariate sorts create portfolios within a two-dimensional space spanned by two sorting variables. It is then possible to assess the return impact of either sorting variable by the return differential from a trading strategy that invests in the portfolios at either end of the respective variables spectrum. We create a five-by-five matrix using book-to-market and firm size as sorting variables in our example below. We end up with 25 portfolios. Since we are interested in the *value premium* (i.e., the return differential between high and low book-to-market firms), we go long the five portfolios of the highest book-to-market firms and short the five portfolios of the lowest book-to-market firms. The five portfolios at each end are due to the size splits we employed alongside the book-to-market splits.

To implement the independent bivariate portfolio sort, we assign monthly portfolios for each of our sorting variables separately to create the variables `portfolio_bm` and `portfolio_me`, respectively. Then, these separate portfolios are combined to the final sort stored in `portfolio_combined`. After assigning the portfolios, we compute the average return within each portfolio for each month. Additionally, we keep the book-to-market portfolio as it makes the computation of the value premium easier. The alternative would be to disaggregate the combined portfolio in a separate step. Notice that we weigh the stocks within each portfolio by their market capitalization, i.e., we decide to value-weight our returns.

```
value_portfolios = (data_for_sorts
  .groupby("month")
  .apply(lambda x: x.assign(
     portfolio_bm=assign_portfolio(
       data=x, sorting_variable="bm", n_portfolios=5, exchanges=["NYSE"]
     ),
     portfolio_me=assign_portfolio(
```

```
      data=x, sorting_variable="me", n_portfolios=5, exchanges=["NYSE"]
    )
  )
)
.reset_index(drop=True)
.groupby(["month", "portfolio_bm", "portfolio_me"])
.apply(lambda x: pd.Series({
    "ret": np.average(x["ret_excess"], weights=x["mktcap_lag"])
  })
)
.reset_index()
)
```

Equipped with our monthly portfolio returns, we are ready to compute the value premium. However, we still have to decide how to invest in the five high and the five low book-to-market portfolios. The most common approach is to weigh these portfolios equally, but this is yet another researcher's choice. Then, we compute the return differential between the high and low book-to-market portfolios and show the average value premium.

```
value_premium = (value_portfolios
  .groupby(["month", "portfolio_bm"])
  .aggregate({"ret": "mean"})
  .reset_index()
  .groupby("month")
  .apply(lambda x: pd.Series({
    "value_premium": (
        x.loc[x["portfolio_bm"] == x["portfolio_bm"].max(), "ret"].mean() -
        x.loc[x["portfolio_bm"] == x["portfolio_bm"].min(), "ret"].mean()
    )
  })
)
  .aggregate({"value_premium": "mean"})
)
```

The resulting monthly value premium is 0.43 percent with an annualized return of 5.3 percent.

## 10.4 Dependent Sorts

In the previous exercise, we assigned the portfolios without considering the second variable in the assignment. This protocol is called independent portfolio sorts. The alternative, i.e., dependent sorts, creates portfolios for the second sorting variable within each bucket of the first sorting variable. In our example below, we sort firms into five size buckets, and within each of those buckets, we assign firms to five book-to-market portfolios. Hence, we have monthly breakpoints that are specific to each size group. The decision between independent and dependent portfolio sorts is another choice for the researcher. Notice that dependent sorts ensure an equal amount of stocks within each portfolio.

To implement the dependent sorts, we first create the size portfolios by calling `assign_portfolio()` with `sorting_variable="me"`. Then, we group our data again by month and by the size portfolio before assigning the book-to-market portfolio. The rest of the implementation is the same as before. Finally, we compute the value premium.

```python
value_portfolios = (data_for_sorts
  .groupby("month")
  .apply(lambda x: x.assign(
      portfolio_me=assign_portfolio(
        data=x, sorting_variable="me", n_portfolios=5, exchanges=["NYSE"]
      )
    )
  )
  .reset_index(drop=True)
  .groupby(["month", "portfolio_me"])
  .apply(lambda x: x.assign(
      portfolio_bm=assign_portfolio(
        data=x, sorting_variable="bm", n_portfolios=5, exchanges=["NYSE"]
      )
    )
  )
  .reset_index(drop=True)
  .groupby(["month", "portfolio_bm", "portfolio_me"])
  .apply(lambda x: pd.Series({
      "ret": np.average(x["ret_excess"], weights=x["mktcap_lag"])
    })
  )
  .reset_index()
)

value_premium = (value_portfolios
  .groupby(["month", "portfolio_bm"])
  .aggregate({"ret": "mean"})
  .reset_index()
  .groupby("month")
  .apply(lambda x: pd.Series({
    "value_premium": (
        x.loc[x["portfolio_bm"] == x["portfolio_bm"].max(), "ret"].mean() -
        x.loc[x["portfolio_bm"] == x["portfolio_bm"].min(), "ret"].mean()
      )
    })
  )
  .aggregate({"value_premium": "mean"})
)
```

The monthly value premium from dependent sorts is 0.38 percent, which translates to an annualized premium of 4.6 percent per year.

Overall, we show how to conduct bivariate portfolio sorts in this chapter. In one case, we sort the portfolios independently of each other. Yet we also discuss how to create dependent portfolio sorts. Along the lines of Chapter 9, we see how many choices a researcher has to make to implement portfolio sorts, and bivariate sorts increase the number of choices.

## 10.5  Exercises

1. In Chapter 9, we examine the distribution of market equity. Repeat this analysis for book equity and the book-to-market ratio (alongside a plot of the breakpoints, i.e., deciles).

2. When we investigate the portfolios, we focus on the returns exclusively. However, it is also of interest to understand the characteristics of the portfolios. Write a function to compute the average characteristics for size and book-to-market across the 25 independently and dependently sorted portfolios.

3. As for the size premium, also the value premium constructed here does not follow Fama and French (1993). Implement a p-hacking setup as in Chapter 9 to find a premium that comes closest to their HML premium.

# 11

## Replicating Fama-French Factors

In this chapter, we provide a replication of the famous Fama-French factor portfolios. The Fama-French factor models are a cornerstone of empirical asset pricing (see Fama and French, 1992, and Fama and French (2015)). On top of the market factor represented by the traditional CAPM beta, the three-factor model includes the size and value factors to explain the cross section of returns. Its successor, the five-factor model, additionally includes profitability and investment as explanatory factors.

We start with the three-factor model. We already introduced the size and value factors in Chapter 10, and their definition remains the same: size is the SMB factor (small-minus-big) that is long small firms and short large firms. The value factor is HML (high-minus-low) and is long in high book-to-market firms and short in low book-to-market counterparts.

After the replication of the three-factor model, we move to the five-factors by constructing the profitability factor RMW (robust-minus-weak) as the difference between the returns of firms with high and low operating profitability and the investment factor CMA (conservative-minus-aggressive) as the difference between firms with high versus low investment rates.

The current chapter relies on this set of Python packages.

```python
import pandas as pd
import numpy as np
import sqlite3
import statsmodels.formula.api as smf

from regtabletotext import prettify_result
```

## 11.1 Data Preparation

We use CRSP and Compustat as data sources, as we need exactly the same variables to compute the size and value factors in the way Fama and French do it. Hence, there is nothing new below, and we only load data from our SQLite database introduced in Chapter 3 and Chapter 4.

```python
tidy_finance = sqlite3.connect(
  database="data/tidy_finance_python.sqlite"
)


crsp_monthly = (pd.read_sql_query(
    sql=("SELECT permno, gvkey, month, ret_excess, mktcap, "
        "mktcap_lag, exchange FROM crsp_monthly"),
```

```
    con=tidy_finance,
    parse_dates={"month"})
  .dropna()
)

compustat = (pd.read_sql_query(
    sql="SELECT gvkey, datadate, be, op, inv FROM compustat",
    con=tidy_finance,
    parse_dates={"datadate"})
  .dropna()
)

factors_ff3_monthly = pd.read_sql_query(
  sql="SELECT month, smb, hml FROM factors_ff3_monthly",
  con=tidy_finance,
  parse_dates={"month"}
)

factors_ff5_monthly = pd.read_sql_query(
  sql="SELECT month, smb, hml, rmw, cma FROM factors_ff5_monthly",
  con=tidy_finance,
  parse_dates={"month"}
)
```

Yet when we start merging our dataset for computing the premiums, there are a few differences to Chapter 10. First, Fama and French form their portfolios in June of year $t$, whereby the returns of July are the first monthly return for the respective portfolio. For firm size, they consequently use the market capitalization recorded for June. It is then held constant until June of year $t + 1$.

Second, Fama and French also have a different protocol for computing the book-to-market ratio. They use market equity as of the end of year $t - 1$ and the book equity reported in year $t - 1$, i.e., the datadate is within the last year. Hence, the book-to-market ratio can be based on accounting information that is up to 18 months old. Market equity also does not necessarily reflect the same time point as book equity.

To implement all these time lags, we again employ the temporary sorting_date-column. Notice that when we combine the information, we want to have a single observation per year and stock since we are only interested in computing the breakpoints held constant for the entire year. We ensure this by a call of drop_duplicates() at the end of the chunk below.

```
size = (crsp_monthly
  .query("month.dt.month == 6")
  .assign(sorting_date=lambda x: (x["month"]+pd.DateOffset(months=1)))
  .get(["permno", "exchange", "sorting_date", "mktcap"])
  .rename(columns={"mktcap": "size"})
)

market_equity = (crsp_monthly
  .query("month.dt.month == 12")
  .assign(sorting_date=lambda x: (x["month"]+pd.DateOffset(months=7)))
```

```
  .get(["permno", "gvkey", "sorting_date", "mktcap"])
  .rename(columns={"mktcap": "me"})
)

book_to_market = (compustat
  .assign(
    sorting_date=lambda x: (pd.to_datetime(
      (x["datadate"].dt.year+1).astype(str)+"0701", format="%Y%m%d")
    )
  )
  .merge(market_equity, how="inner", on=["gvkey", "sorting_date"])
  .assign(bm=lambda x: x["be"]/x["me"])
  .get(["permno", "sorting_date", "me", "bm"])
)

sorting_variables = (size
  .merge(book_to_market, how="inner", on=["permno", "sorting_date"])
  .dropna()
  .drop_duplicates(subset=["permno", "sorting_date"])
)
```

## 11.2 Portfolio Sorts

Next, we construct our portfolios with an adjusted `assign_portfolio()` function. Fama and French rely on NYSE-specific breakpoints, they form two portfolios in the size dimension at the median and three portfolios in the dimension of book-to-market at the 30- and 70-percentiles, and they use independent sorts. The sorts for book-to-market require an adjustment to the function in Chapter 10 because it would not produce the right breakpoints. Instead of `n_portfolios`, we now specify `percentiles`, which take the breakpoint-sequence as an object specified in the function's call. Specifically, we give `percentiles = [0, 0.3, 0.7, 1]` to the function. Additionally, we perform a join with our return data to ensure that we only use traded stocks when computing the breakpoints as a first step.

```
def assign_portfolio(data, sorting_variable, percentiles):
    """Assign portfolios to a bin according to a sorting variable."""

    breakpoints = (data
      .query("exchange == 'NYSE'")
      .get(sorting_variable)
      .quantile(percentiles, interpolation="linear")
    )
    breakpoints.iloc[0] = -np.Inf
    breakpoints.iloc[breakpoints.size-1] = np.Inf

    assigned_portfolios = pd.cut(
      data[sorting_variable],
      bins=breakpoints,
```

```
        labels=pd.Series(range(1, breakpoints.size)),
        include_lowest=True,
        right=False
    )

    return assigned_portfolios

portfolios = (sorting_variables
  .groupby("sorting_date")
  .apply(lambda x: x
    .assign(
      portfolio_size=assign_portfolio(x, "size", [0, 0.5, 1]),
      portfolio_bm=assign_portfolio(x, "bm", [0, 0.3, 0.7, 1])
    )
  )
  .reset_index(drop=True)
  .get(["permno", "sorting_date", "portfolio_size", "portfolio_bm"])
)
```

Next, we merge the portfolios to the return data for the rest of the year. To implement this step, we create a new column `sorting_date` in our return data by setting the date to sort on to July of $t-1$ if the month is June (of year $t$) or earlier or to July of year $t$ if the month is July or later.

```
portfolios = (crsp_monthly
  .assign(
    sorting_date=lambda x: (pd.to_datetime(
      x["month"].apply(lambda x: str(x.year-1)+
        "0701" if x.month <= 6 else str(x.year)+"0701")))
  )
  .merge(portfolios, how="inner", on=["permno", "sorting_date"])
)
```

## 11.3   Fama-French Three-Factor Model

Equipped with the return data and the assigned portfolios, we can now compute the value-weighted average return for each of the six portfolios. Then, we form the Fama-French factors. For the size factor (i.e., SMB), we go long in the three small portfolios and short the three large portfolios by taking an average across either group. For the value factor (i.e., HML), we go long in the two high book-to-market portfolios and short the two low book-to-market portfolios, again weighting them equally (using the `mean()` function).

```
factors_replicated = (portfolios
  .groupby(["portfolio_size", "portfolio_bm", "month"])
  .apply(lambda x: pd.Series({
    "ret": np.average(x["ret_excess"], weights=x["mktcap_lag"])
    })
  )
```

```
  .reset_index()
  .groupby("month")
  .apply(lambda x: pd.Series({
    "smb_replicated": (
      x["ret"][x["portfolio_size"] == 1].mean() -
        x["ret"][x["portfolio_size"] == 2].mean()),
    "hml_replicated": (
      x["ret"][x["portfolio_bm"] == 3].mean() -
        x["ret"][x["portfolio_bm"] == 1].mean())
    }))
  .reset_index()
)

factors_replicated = (factors_replicated
  .merge(factors_ff3_monthly, how="inner", on="month")
  .round(4)
)
```

## 11.4   Replication Evaluation

In the previous section, we replicated the size and value premiums following the procedure outlined by Fama and French. The final question is then: how close did we get? We answer this question by looking at the two time-series estimates in a regression analysis using smf.ols(). If we did a good job, then we should see a non-significant intercept (rejecting the notion of systematic error), a coefficient close to one (indicating a high correlation), and an adjusted R-squared close to one (indicating a high proportion of explained variance).

To test the success of the SMB factor, we hence run the following regression:

```
model_smb = (smf.ols(
    formula="smb ~ smb_replicated",
    data=factors_replicated
  )
  .fit()
)
prettify_result(model_smb)
```

```
OLS Model:
smb ~ smb_replicated

Coefficients:
                Estimate  Std. Error  t-Statistic  p-Value
Intercept         -0.000       0.000       -1.334    0.183
smb_replicated     0.994       0.004      231.343    0.000

Summary statistics:
- Number of observations: 726
- R-squared: 0.987, Adjusted R-squared: 0.987
```

- F-statistic: 53,519.615 on 1 and 724 DF, p-value: 0.000

The results for the SMB factor are quite convincing as all three criteria outlined above are met and the coefficient is 0.99 and R-squared are at 99 percent.

```
model_hml = (smf.ols(
    formula="hml ~ hml_replicated",
    data=factors_replicated
  )
  .fit()
)
prettify_result(model_hml)
```

```
OLS Model:
hml ~ hml_replicated

Coefficients:
                Estimate  Std. Error  t-Statistic  p-Value
Intercept          0.000       0.000        1.633    0.103
hml_replicated     0.963       0.007      133.678    0.000

Summary statistics:
- Number of observations: 726
- R-squared: 0.961, Adjusted R-squared: 0.961
- F-statistic: 17,869.727 on 1 and 724 DF, p-value: 0.000
```

The replication of the HML factor is also a success, although at a slightly lower level with coefficient of 0.96 and R-squared around 96 percent.

The evidence allows us to conclude that we did a relatively good job in replicating the original Fama-French size and value premiums, although we do not know their underlying code. From our perspective, a perfect match is only possible with additional information from the maintainers of the original data.

## 11.5   Fama-French Five-Factor Model

Now, let us move to the replication of the five-factor model. We extend the `other_sorting_variables` table from above with the additional characteristics operating profitability op and investment inv. Note that the `dropna()` statement yields different sample sizes, as some firms with be values might not have op or inv values.

```
other_sorting_variables = (compustat
  .assign(
    sorting_date=lambda x: (pd.to_datetime(
      (x["datadate"].dt.year+1).astype(str)+"0701", format="%Y%m%d")
    )
  )
  .merge(market_equity, how="inner", on=["gvkey", "sorting_date"])
  .assign(bm=lambda x: x["be"]/x["me"])
  .get(["permno", "sorting_date", "me", "bm", "op", "inv"])
```

```
)

sorting_variables = (size
  .merge(other_sorting_variables, how="inner", on=["permno", "sorting_date"])
  .dropna()
  .drop_duplicates(subset=["permno", "sorting_date"])
)
```

In each month, we independently sort all stocks into the two size portfolios. The value, profitability, and investment portfolios, on the other hand, are the results of dependent sorts based on the size portfolios. We then merge the portfolios to the return data for the rest of the year just as above.

```
portfolios = (sorting_variables
  .groupby("sorting_date")
  .apply(lambda x: x
    .assign(
      portfolio_size=assign_portfolio(x, "size", [0, 0.5, 1])
    )
  )
  .reset_index(drop=True)
  .groupby(["sorting_date", "portfolio_size"])
  .apply(lambda x: x
    .assign(
      portfolio_bm=assign_portfolio(x, "bm", [0, 0.3, 0.7, 1]),
      portfolio_op=assign_portfolio(x, "op", [0, 0.3, 0.7, 1]),
      portfolio_inv=assign_portfolio(x, "inv", [0, 0.3, 0.7, 1])
    )
  )
  .reset_index(drop=True)
  .get(["permno", "sorting_date",
        "portfolio_size", "portfolio_bm",
        "portfolio_op", "portfolio_inv"])
)

portfolios = (crsp_monthly
  .assign(
    sorting_date=lambda x: (pd.to_datetime(
      x["month"].apply(lambda x: str(x.year-1)+
        "0701" if x.month <= 6 else str(x.year)+"0701")))
  )
  .merge(portfolios, how="inner", on=["permno", "sorting_date"])
)
```

Now, we want to construct each of the factors, but this time, the size factor actually comes last because it is the result of averaging across all other factor portfolios. This dependency is the reason why we keep the table with value-weighted portfolio returns as a separate object that we reuse later. We construct the value factor, HML, as above by going long the two portfolios with high book-to-market ratios and shorting the two portfolios with low book-to-market.

```
portfolios_value = (portfolios
  .groupby(["portfolio_size", "portfolio_bm", "month"])
  .apply(lambda x: pd.Series({
      "ret": np.average(x["ret_excess"], weights=x["mktcap_lag"])
    })
  )
  .reset_index()
)

factors_value = (portfolios_value
  .groupby("month")
  .apply(lambda x: pd.Series({
    "hml_replicated": (
      x["ret"][x["portfolio_bm"] == 3].mean() -
        x["ret"][x["portfolio_bm"] == 1].mean())})
  )
  .reset_index()
)
```

For the profitability factor, RMW (robust-minus-weak), we take a long position in the two high profitability portfolios and a short position in the two low profitability portfolios.

```
portfolios_profitability = (portfolios
  .groupby(["portfolio_size", "portfolio_op", "month"])
  .apply(lambda x: pd.Series({
      "ret": np.average(x["ret_excess"], weights=x["mktcap_lag"])
    })
  )
  .reset_index()
)

factors_profitability = (portfolios_profitability
  .groupby("month")
  .apply(lambda x: pd.Series({
    "rmw_replicated": (
      x["ret"][x["portfolio_op"] == 3].mean() -
        x["ret"][x["portfolio_op"] == 1].mean())})
  )
  .reset_index()
)
```

For the investment factor, CMA (conservative-minus-aggressive), we go long the two low investment portfolios and short the two high investment portfolios.

```
portfolios_investment = (portfolios
  .groupby(["portfolio_size", "portfolio_inv", "month"])
  .apply(lambda x: pd.Series({
      "ret": np.average(x["ret_excess"], weights=x["mktcap_lag"])
    })
  )
  .reset_index()
)
```

```
factors_investment = (portfolios_investment
  .groupby("month")
  .apply(lambda x: pd.Series({
    "cma_replicated": (
      x["ret"][x["portfolio_inv"] == 1].mean() -
        x["ret"][x["portfolio_inv"] == 3].mean())})
  )
  .reset_index()
)
```

Finally, the size factor, SMB, is constructed by going long the six small portfolios and short the six large portfolios.

```
factors_size = (
  pd.concat(
    [portfolios_value, portfolios_profitability, portfolios_investment],
    ignore_index=True
  )
  .groupby("month")
  .apply(lambda x: pd.Series({
    "smb_replicated": (
      x["ret"][x["portfolio_size"] == 1].mean() -
        x["ret"][x["portfolio_size"] == 2].mean())})
  )
  .reset_index()
)
```

We then join all factors together into one dataframe and construct again a suitable table to run tests for evaluating our replication.

```
factors_replicated = (factors_size
  .merge(factors_value, how="outer", on="month")
  .merge(factors_profitability, how="outer", on="month")
  .merge(factors_investment, how="outer", on="month")
)

factors_replicated = (factors_replicated
  .merge(factors_ff5_monthly, how="inner", on="month")
  .round(4)
)
```

Let us start the replication evaluation again with the size factor:

```
model_smb = (smf.ols(
    formula="smb ~ smb_replicated",
    data=factors_replicated
  )
  .fit()
)
prettify_result(model_smb)

OLS Model:
smb ~ smb_replicated
```

```
Coefficients:
                Estimate  Std. Error  t-Statistic  p-Value
Intercept         -0.00       0.000       -1.493    0.136
smb_replicated     0.97       0.004      222.489    0.000

Summary statistics:
- Number of observations: 714
- R-squared: 0.986, Adjusted R-squared: 0.986
- F-statistic: 49,501.421 on 1 and 712 DF, p-value: 0.000
```

The results for the SMB factor are quite convincing, as all three criteria outlined above are met and the coefficient is 0.97 and the R-squared is at 99 percent.

```
model_hml = (smf.ols(
    formula="hml ~ hml_replicated",
    data=factors_replicated
  )
  .fit()
)
prettify_result(model_hml)
```

```
OLS Model:
hml ~ hml_replicated

Coefficients:
                Estimate  Std. Error  t-Statistic  p-Value
Intercept          0.000        0.00        1.588    0.113
hml_replicated     0.991        0.01       96.694    0.000

Summary statistics:
- Number of observations: 714
- R-squared: 0.929, Adjusted R-squared: 0.929
- F-statistic: 9,349.714 on 1 and 712 DF, p-value: 0.000
```

The replication of the HML factor is also a success, although at a slightly higher coefficient of 0.99 and an R-squared around 93 percent.

```
model_rmw = (smf.ols(
    formula="rmw ~ rmw_replicated",
    data=factors_replicated
  )
  .fit()
)
prettify_result(model_rmw)
```

```
OLS Model:
rmw ~ rmw_replicated

Coefficients:
                Estimate  Std. Error  t-Statistic  p-Value
Intercept          0.000       0.000        0.268    0.789
rmw_replicated     0.954       0.009      107.038    0.000
```

Summary statistics:
- Number of observations: 714
- R-squared: 0.941, Adjusted R-squared: 0.941
- F-statistic: 11,457.150 on 1 and 712 DF, p-value: 0.000

We are also able to replicate the RMW factor quite well with a coefficient of 0.95 and an R-squared around 94 percent.

```
model_cma = (smf.ols(
    formula="cma ~ cma_replicated",
    data=factors_replicated
  )
  .fit()
)
prettify_result(model_cma)
```

OLS Model:
cma ~ cma_replicated

Coefficients:
|                | Estimate | Std. Error | t-Statistic | p-Value |
|----------------|----------|------------|-------------|---------|
| Intercept      | 0.001    | 0.000      | 4.002       | 0.0     |
| cma_replicated | 0.964    | 0.008      | 117.396     | 0.0     |

Summary statistics:
- Number of observations: 714
- R-squared: 0.951, Adjusted R-squared: 0.951
- F-statistic: 13,781.776 on 1 and 712 DF, p-value: 0.000

Finally, the CMA factor also replicates well with a coefficient of 0.96 and an R-squared around 95 percent.

Overall, our approach seems to replicate the Fama-French three and five-factor models just as well as the three-factors.

## 11.6 Exercises

1. Fama and French (1993) claim that their sample excludes firms until they have appeared in Compustat for two years. Implement this additional filter and compare the improvements of your replication effort.
2. On his homepage, Kenneth French[1] provides instructions on how to construct the most common variables used for portfolio sorts. Try to replicate the univariate portfolio sort return time series for E/P (earnings/price) provided on his homepage and evaluate your replication effort using regressions.

---

[1] https://mba.tuck.dartmouth.edu/pages/faculty/ken.french/Data_Library/variable_definitions.html

# 12

# Fama-MacBeth Regressions

In this chapter, we present a simple implementation of Fama and MacBeth (1973), a regression approach commonly called Fama-MacBeth regressions. Fama-MacBeth regressions are widely used in empirical asset pricing studies. We use individual stocks as test assets to estimate the risk premium associated with the three factors included in Fama and French (1993).

Researchers use the two-stage regression approach to estimate risk premiums in various markets, but predominately in the stock market. Essentially, the two-step Fama-MacBeth regressions exploit a linear relationship between expected returns and exposure to (priced) risk factors. The basic idea of the regression approach is to project asset returns on factor exposures or characteristics that resemble exposure to a risk factor in the cross-section in each time period. Then, in the second step, the estimates are aggregated across time to test if a risk factor is priced. In principle, Fama-MacBeth regressions can be used in the same way as portfolio sorts introduced in previous chapters.

The Fama-MacBeth procedure is a simple two-step approach: The first step uses the exposures (characteristics) as explanatory variables in $T$ cross-sectional regressions. For example, if $r_{i,t+1}$ denote the excess returns of asset $i$ in month $t+1$, then the famous Fama-French three-factor model implies the following return generating process (see also Campbell et al., 1998):

$$r_{i,t+1} = \alpha_i + \lambda_t^M \beta_{i,t}^M + \lambda_t^{SMB} \beta_{i,t}^{SMB} + \lambda_t^{HML} \beta_{i,t}^{HML} + \epsilon_{i,t}. \qquad (12.1)$$

Here, we are interested in the compensation $\lambda_t^f$ for the exposure to each risk factor $\beta_{i,t}^f$ at each time point, i.e., the risk premium. Note the terminology: $\beta_{i,t}^f$ is an asset-specific characteristic, e.g., a factor exposure or an accounting variable. *If* there is a linear relationship between expected returns and the characteristic in a given month, we expect the regression coefficient to reflect the relationship, i.e., $\lambda_t^f \neq 0$.

In the second step, the time-series average $\frac{1}{T} \sum_{t=1}^{T} \hat{\lambda}_t^f$ of the estimates $\hat{\lambda}_t^f$ can then be interpreted as the risk premium for the specific risk factor $f$. We follow Zaffaroni and Zhou (2022) and consider the standard cross-sectional regression to predict future returns. If the characteristics are replaced with time $t+1$ variables, then the regression approach captures risk attributes rather than risk premiums.

Before we move to the implementation, we want to highlight that the characteristics, e.g., $\hat{\beta}_i^f$, are often estimated in a separate step before applying the actual Fama-MacBeth methodology. You can think of this as a *step 0*. You might thus worry that the errors of $\hat{\beta}_i^f$ impact the risk premiums' standard errors. Measurement error in $\hat{\beta}_i^f$ indeed affects the risk premium estimates, i.e., they lead to biased estimates. The literature provides adjustments for this bias (see, e.g., Shanken, 1992; Kim, 1995; Chen et al., 2015, among others) but also shows that the bias goes to zero as $T \to \infty$. We refer to Gagliardini et al. (2016) for an in-depth discussion also covering the case of time-varying betas. Moreover, if you plan to use Fama-MacBeth regressions with individual stocks, Hou et al. (2020) advocates using weighted-least squares to estimate the coefficients such that they are not biased toward small firms. Without this adjustment, the high number of small firms would drive the coefficient estimates.

The current chapter relies on this set of Python packages.

```python
import pandas as pd
import numpy as np
import sqlite3
import statsmodels.formula.api as smf
```

## 12.1   Data Preparation

We illustrate Fama and MacBeth (1973) with the monthly CRSP sample and use three characteristics to explain the cross-section of returns: Market capitalization, the book-to-market ratio, and the CAPM beta (i.e., the covariance of the excess stock returns with the market excess returns). We collect the data from our SQLite database introduced in Chapter 3 and Chapter 4.

```python
tidy_finance = sqlite3.connect(database="data/tidy_finance_python.sqlite")

crsp_monthly = pd.read_sql_query(
  sql="SELECT permno, gvkey, month, ret_excess, mktcap FROM crsp_monthly",
  con=tidy_finance,
  parse_dates={"month"}
)

compustat = pd.read_sql_query(
  sql="SELECT datadate, gvkey, be FROM compustat",
  con=tidy_finance,
  parse_dates={"datadate"}
)

beta = pd.read_sql_query(
  sql="SELECT month, permno, beta_monthly FROM beta",
  con=tidy_finance,
  parse_dates={"month"}
)
```

We use the Compustat and CRSP data to compute the book-to-market ratio and the (log) market capitalization. Furthermore, we also use the CAPM betas based on monthly returns we computed in the previous chapters.

```python
characteristics = (compustat
  .assign(month=lambda x: x["datadate"].dt.to_period("M").dt.to_timestamp())
  .merge(crsp_monthly, how="left", on=["gvkey", "month"], )
  .merge(beta, how="left", on=["permno", "month"])
  .assign(
    bm=lambda x: x["be"]/x["mktcap"],
```

```
      log_mktcap=lambda x: np.log(x["mktcap"]),
      sorting_date=lambda x: x["month"]+pd.DateOffset(months=6)
  )
  .get(["gvkey", "bm", "log_mktcap", "beta_monthly", "sorting_date"])
  .rename(columns={"beta_monthly": "beta"})
)

data_fama_macbeth = (crsp_monthly
  .merge(characteristics,
         how="left",
         left_on=["gvkey", "month"], right_on=["gvkey", "sorting_date"])
  .sort_values(["month", "permno"])
  .groupby("permno")
  .apply(lambda x: x.assign(
      beta=x["beta"].fillna(method="ffill"),
      bm=x["bm"].fillna(method="ffill"),
      log_mktcap=x["log_mktcap"].fillna(method="ffill")
    )
  )
  .reset_index(drop=True)
)

data_fama_macbeth_lagged = (data_fama_macbeth
  .assign(month=lambda x: x["month"]-pd.DateOffset(months=1))
  .get(["permno", "month", "ret_excess"])
  .rename(columns={"ret_excess": "ret_excess_lead"})
)

data_fama_macbeth = (data_fama_macbeth
  .merge(data_fama_macbeth_lagged, how="left", on=["permno", "month"])
  .get(["permno", "month", "ret_excess_lead", "beta", "log_mktcap", "bm"])
  .dropna()
)
```

## 12.2   Cross-Sectional Regression

Next, we run the cross-sectional regressions with the characteristics as explanatory variables for each month. We regress the returns of the test assets at a particular time point on the characteristics of each asset. By doing so, we get an estimate of the risk premiums $\hat{\lambda}_t^f$ for each point in time.

```
risk_premiums = (data_fama_macbeth
  .groupby("month")
  .apply(lambda x: smf.ols(
      formula="ret_excess_lead ~ beta + log_mktcap + bm",
      data=x
    ).fit()
```

```
      .params
  )
  .reset_index()
)
```

## 12.3  Time-Series Aggregation

Now that we have the risk premiums' estimates for each period, we can average across the time-series dimension to get the expected risk premium for each characteristic. Similarly, we manually create the *t*-test statistics for each regressor, which we can then compare to usual critical values of 1.96 or 2.576 for two-tailed significance tests at a five percent and a one percent significance level.

```
price_of_risk = (risk_premiums
  .melt(id_vars="month", var_name="factor", value_name="estimate")
  .groupby("factor")["estimate"]
  .apply(lambda x: pd.Series({
      "risk_premium": 100*x.mean(),
      "t_statistic": x.mean()/x.std()*np.sqrt(len(x))
    })
  )
  .reset_index()
  .pivot(index="factor", columns="level_1", values="estimate")
  .reset_index()
)
```

It is common to adjust for autocorrelation when reporting standard errors of risk premiums. As in Chapter 8, the typical procedure for this is computing Newey and West (1987) standard errors.

```
price_of_risk_newey_west = (risk_premiums
  .melt(id_vars="month", var_name="factor", value_name="estimate")
  .groupby("factor")
  .apply(lambda x: (
      x["estimate"].mean()/
        smf.ols("estimate ~ 1", x)
        .fit(cov_type="HAC", cov_kwds={"maxlags": 6}).bse
    )
  )
  .reset_index()
  .rename(columns={"Intercept": "t_statistic_newey_west"})
)

(price_of_risk
  .merge(price_of_risk_newey_west, on="factor")
  .round(3)
)
```

| | factor | risk_premium | t_statistic | t_statistic_newey_west |
|---|---|---|---|---|
| 0 | Intercept | 1.244 | 4.809 | 4.167 |
| 1 | beta | 0.001 | 0.013 | 0.012 |
| 2 | bm | 0.161 | 3.054 | 2.860 |
| 3 | log_mktcap | -0.108 | -3.004 | -2.828 |

Finally, let us interpret the results. Stocks with higher book-to-market ratios earn higher expected future returns, which is in line with the value premium. The negative value for log market capitalization reflects the size premium for smaller stocks. Consistent with results from earlier chapters, we detect no relation between beta and future stock returns.

## 12.4 Exercises

1. Download a sample of test assets from Kenneth French's homepage and reevaluate the risk premiums for industry portfolios instead of individual stocks.
2. Use individual stocks with weighted-least squares based on a firm's size as suggested by Hou et al. (2020). Then, repeat the Fama-MacBeth regressions without the weighting-scheme adjustment but drop the smallest 20 percent of firms each month. Compare the results of the three approaches.

# Part IV

# Modeling and Machine Learning

# 13

# *Fixed Effects and Clustered Standard Errors*

In this chapter, we provide an intuitive introduction to the two popular concepts of *fixed effects regressions* and *clustered standard errors*. When working with regressions in empirical finance, you will sooner or later be confronted with discussions around how you deal with omitted variables bias and dependence in your residuals. The concepts we introduce in this chapter are designed to address such concerns.

We focus on a classical panel regression common to the corporate finance literature (e.g., Fazzari et al., 1988; Erickson and Whited, 2012; Gulen and Ion, 2015): firm investment modeled as a function that increases in firm cash flow and firm investment opportunities.

Typically, this investment regression uses quarterly balance sheet data provided via Compustat because it allows for richer dynamics in the regressors and more opportunities to construct variables. As we focus on the implementation of fixed effects and clustered standard errors, we use the annual Compustat data from our previous chapters and leave the estimation using quarterly data as an exercise. We demonstrate below that the regression based on annual data yields qualitatively similar results to estimations based on quarterly data from the literature, namely confirming the positive relationships between investment and the two regressors.

The current chapter relies on the following set of Python packages.

```
import pandas as pd
import numpy as np
import sqlite3
import datetime as dt
import itertools
import linearmodels as lm

from regtabletotext import prettify_result, prettify_result
```

Compared to previous chapters, we introduce `linearmodels` (Sheppard, 2023), which provides tools for estimating various econometric models such as panel regressions.

## 13.1 Data Preparation

We use CRSP and annual Compustat as data sources from our SQLite database introduced in Chapter 3 and Chapter 4. In particular, Compustat provides balance sheet and income statement data on a firm level, while CRSP provides market valuations.

```
tidy_finance = sqlite3.connect(
  database="data/tidy_finance_python.sqlite"
)


crsp_monthly = pd.read_sql_query(
  sql="SELECT gvkey, month, mktcap FROM crsp_monthly",
  con=tidy_finance,
  parse_dates={"month"}
)


compustat = pd.read_sql_query(
  sql=("SELECT datadate, gvkey, year, at, be, capx, oancf, txdb "
       "FROM compustat"),
  con=tidy_finance,
  parse_dates={"datadate"}
)
```

The classical investment regressions model is the capital investment of a firm as a function of operating cash flows and Tobin's q, a measure of a firm's investment opportunities. We start by constructing investment and cash flows which are usually normalized by lagged total assets of a firm. In the following code chunk, we construct a *panel* of firm-year observations, so we have both cross-sectional information on firms as well as time-series information for each firm.

```
data_investment = (compustat
  .assign(
    month=lambda x: (
      pd.to_datetime(x["datadate"]).dt.to_period("M").dt.to_timestamp()
    )
  )
  .merge(compustat.get(["gvkey", "year", "at"])
         .rename(columns={"at": "at_lag"})
         .assign(year=lambda x: x["year"]+1),
         on=["gvkey", "year"], how="left")
  .query("at > 0 and at_lag > 0")
  .assign(investment=lambda x: x["capx"]/x["at_lag"],
          cash_flows=lambda x: x["oancf"]/x["at_lag"])
)


data_investment = (data_investment
  .merge(data_investment.get(["gvkey", "year", "investment"])
         .rename(columns={"investment": "investment_lead"})
         .assign(year=lambda x: x["year"]-1),
         on=["gvkey", "year"], how="left")
)
```

Tobin's q is the ratio of the market value of capital to its replacement costs. It is one of the most common regressors in corporate finance applications (e.g., Fazzari et al., 1988; Erickson and Whited, 2012). We follow the implementation of Gulen and Ion (2015) and compute Tobin's q as the market value of equity (`mktcap`) plus the book value of assets (`at`) minus book value of equity (`be`) plus deferred taxes (`txdb`), all divided by book value of assets (`at`). Finally, we only keep observations where all variables of interest are non-missing, and the reported book value of assets is strictly positive.

```
data_investment = (data_investment
  .merge(crsp_monthly, on=["gvkey", "month"], how="left")
  .assign(
    tobins_q=lambda x: (
      (x["mktcap"]+x["at"]-x["be"]+x["txdb"])/x["at"]
    )
  )
  .get(["gvkey", "year", "investment_lead", "cash_flows", "tobins_q"])
  .dropna()
)
```

As the variable construction typically leads to extreme values that are most likely related to data issues (e.g., reporting errors), many papers include winsorization of the variables of interest. Winsorization involves replacing values of extreme outliers with quantiles on the respective end. The following function implements the winsorization for any percentage cut that should be applied on either end of the distributions. In the specific example, we winsorize the main variables (`investment`, `cash_flows`, and `tobins_q`) at the one percent level.[1]

```
def winsorize(x, cut):
    """Winsorize returns at cut level."""

    tmp_x = x.copy()
    upper_quantile=np.nanquantile(tmp_x, 1 - cut)
    lower_quantile=np.nanquantile(tmp_x, cut)
    tmp_x[tmp_x > upper_quantile]=upper_quantile
    tmp_x[tmp_x < lower_quantile]=lower_quantile

    return tmp_x

data_investment = (data_investment
  .assign(
    investment_lead=lambda x: winsorize(x["investment_lead"], 0.01),
    cash_flows=lambda x: winsorize(x["cash_flows"], 0.01),
    tobins_q=lambda x: winsorize(x["tobins_q"], 0.01)
  )
)
```

Before proceeding to any estimations, we highly recommend tabulating summary statistics of the variables that enter the regression. These simple tables allow you to check the plausibility of your numerical variables, as well as spot any obvious errors or outliers. Additionally, for panel data, plotting the time series of the variable's mean and the number of observations is a useful exercise to spot potential problems.

```
data_investment_summary = (data_investment
  .melt(id_vars=["gvkey", "year"], var_name="measure",
```

---

[1] Note that in `pandas`, when you index a dataframee, you receive a reference to the original dataframee. Consequently, modifying a subset will directly impact the initial dataframee. To prevent unintended changes to the original dataframee, it is advisable to use the `copy()` method as we do here in the `winsorize` function.

```
        value_vars=["investment_lead", "cash_flows", "tobins_q"])
    .get(["measure", "value"])
    .groupby("measure")
    .describe(percentiles=[0.05, 0.5, 0.95])
)
np.round(data_investment_summary, 2)
```

|  | value count | mean | std | min | 5% | 50% | 95% | max |
|---|---|---|---|---|---|---|---|---|
| measure |  |  |  |  |  |  |  |  |
| cash_flows | 127468.0 | 0.01 | 0.27 | -1.56 | -0.47 | 0.06 | 0.27 | 0.48 |
| investment_lead | 127468.0 | 0.06 | 0.08 | 0.00 | 0.00 | 0.03 | 0.21 | 0.46 |
| tobins_q | 127468.0 | 2.00 | 1.70 | 0.57 | 0.79 | 1.39 | 5.37 | 10.91 |

## 13.2   Fixed Effects

To illustrate fixed effects regressions, we use the `linearmodels` package, which is both computationally powerful and flexible with respect to model specifications. We start out with the basic investment regression using the simple model

$$\text{Investment}_{i,t+1} = \alpha + \beta_1 \text{Cash Flows}_{i,t} + \beta_2 \text{Tobin's q}_{i,t} + \varepsilon_{i,t}, \qquad (13.1)$$

where $\varepsilon_t$ is i.i.d. normally distributed across time and firms. We use the `PanelOLS()`-function to estimate the simple model so that the output has the same structure as the other regressions below.

```
model_ols = lm.PanelOLS.from_formula(
  formula="investment_lead ~ cash_flows + tobins_q + 1",
  data=data_investment.set_index(["gvkey", "year"]),
).fit()
prettify_result(model_ols)
```

```
Panel OLS Model:
investment_lead ~ cash_flows + tobins_q + 1

Covariance Type: Unadjusted

Coefficients:
            Estimate  Std. Error  t-Statistic  p-Value
Intercept      0.042       0.000      127.470      0.0
cash_flows     0.049       0.001       61.787      0.0
tobins_q       0.007       0.000       57.099      0.0

Summary statistics:
- Number of observations: 127,468
- R-squared (incl. FE): 0.043, Within R-squared: 0.039
```

As expected, the regression output shows significant coefficients for both variables. Higher cash flows and investment opportunities are associated with higher investment. However, the simple model actually may have a lot of omitted variables, so our coefficients are most likely biased. As there is a lot of unexplained variation in our simple model (indicated by the rather low adjusted R-squared), the bias in our coefficients is potentially severe, and the true values could be above or below zero. Note that there are no clear cutoffs to decide when an R-squared is high or low, but it depends on the context of your application and on the comparison of different models for the same data.

One way to tackle the issue of omitted variable bias is to get rid of as much unexplained variation as possible by including *fixed effects*; i.e., model parameters that are fixed for specific groups (e.g., Wooldridge, 2010). In essence, each group has its own mean in fixed effects regressions. The simplest group that we can form in the investment regression is the firm level. The firm fixed effects regression is then

$$\text{Investment}_{i,t+1} = \alpha_i + \beta_1 \text{Cash Flows}_{i,t} + \beta_2 \text{Tobin's q}_{i,t} + \varepsilon_{i,t}, \qquad (13.2)$$

where $\alpha_i$ is the firm fixed effect and captures the firm-specific mean investment across all years. In fact, you could also compute firms' investments as deviations from the firms' average investments and estimate the model without the fixed effects. The idea of the firm fixed effect is to remove the firm's average investment, which might be affected by firm-specific variables that you do not observe. For example, firms in a specific industry might invest more on average. Or you observe a young firm with large investments but only small concurrent cash flows, which will only happen in a few years. This sort of variation is unwanted because it is related to unobserved variables that can bias your estimates in any direction.

To include the firm fixed effect, we use `gvkey` (Compustat's firm identifier) as follows:

```
model_fe_firm = lm.PanelOLS.from_formula(
    formula="investment_lead ~ cash_flows + tobins_q + EntityEffects",
    data=data_investment.set_index(["gvkey", "year"]),
).fit()
prettify_result(model_fe_firm)
```

```
Panel OLS Model:
investment_lead ~ cash_flows + tobins_q + EntityEffects

Covariance Type: Unadjusted

Coefficients:
            Estimate  Std. Error  t-Statistic  p-Value
cash_flows   0.014      0.001       15.219       0.0
tobins_q     0.011      0.000       82.098       0.0

Included Fixed Effects:
         Total
Entity   14350

Summary statistics:
- Number of observations: 127,468
- R-squared (incl. FE): 0.585, Within R-squared: 0.057
```

The regression output shows a lot of unexplained variation at the firm level that is taken care of by including the firm fixed effect as the adjusted R-squared rises above 50 percent. In fact, it is more interesting to look at the within R-squared that shows the explanatory power of a firm's cash flow and Tobin's q *on top* of the average investment of each firm. We can also see that the coefficients changed slightly in magnitude but not in sign.

There is another source of variation that we can get rid of in our setting: average investment across firms might vary over time due to macroeconomic factors that affect all firms, such as economic crises. By including year fixed effects, we can take out the effect of unobservables that vary over time. The two-way fixed effects regression is then

$$\text{Investment}_{i,t+1} = \alpha_i + \alpha_t + \beta_1 \text{Cash Flows}_{i,t} + \beta_2 \text{Tobin's q}_{i,t} + \varepsilon_{i,t}, \qquad (13.3)$$

where $\alpha_t$ is the time fixed effect. Here you can think of higher investments during an economic expansion with simultaneously high cash flows. You can include a time fixed effects by using "TimeEffects" in the formula of PanelOLS.

```
model_fe_firmyear = lm.PanelOLS.from_formula(
    formula=("investment_lead ~ cash_flows + tobins_q + EntityEffects"
            " + TimeEffects"),
    data=data_investment.set_index(["gvkey", "year"]),
).fit()
prettify_result(model_fe_firmyear)
```

```
Panel OLS Model:
investment_lead ~ cash_flows + tobins_q + EntityEffects + TimeEffects

Covariance Type: Unadjusted

Coefficients:
              Estimate  Std. Error  t-Statistic  p-Value
cash_flows      0.018       0.001       19.361      0.0
tobins_q        0.010       0.000       75.419      0.0

Included Fixed Effects:
         Total
Entity   14350
Time        35

Summary statistics:
- Number of observations: 127,468
- R-squared (incl. FE): 0.606, Within R-squared: 0.057
```

The inclusion of time fixed effects did only marginally affect the R-squared and the coefficients, which we can interpret as a good thing as it indicates that the coefficients are not driven by an omitted variable that varies over time.

How can we further improve the robustness of our regression results? Ideally, we want to get rid of unexplained variation at the firm-year level, which means we need to include more variables that vary across firm *and* time and are likely correlated with investment. Note that we cannot include firm-year fixed effects in our setting because then cash flows and Tobin's q are colinear with the fixed effects, and the estimation becomes void.

Before we discuss the properties of our estimation errors, we want to point out that regression tables are at the heart of every empirical analysis, where you compare multiple models. Fortunately, the `results.compare()` function provides a convenient way to tabulate the regression output (with many parameters to customize and even print the output in LaTeX).

We recommend printing *t*-statistics rather than standard errors in regression tables because the latter are typically very hard to interpret across coefficients that vary in size. We also do not print p-values because they are sometimes misinterpreted to signal the importance of observed effects (Wasserstein and Lazar, 2016). The *t*-statistics provide a consistent way to interpret changes in estimation uncertainty across different model specifications.

```
prettify_result([model_ols, model_fe_firm, model_fe_firmyear])
```

| Dependent var. | investment_lead | investment_lead | investment_lead |
|---|---|---|---|
| | | | |
| Intercept | 0.042 (127.47) | | |
| cash_flows | 0.049 (61.79) | 0.014 (15.22) | 0.018 (19.36) |
| tobins_q | 0.007 (57.1) | 0.011 (82.1) | 0.01 (75.42) |
| | | | |
| Fixed effects | | Entity | Entity, Time |
| VCOV type | Unadjusted | Unadjusted | Unadjusted |
| Observations | 127,468 | 127,468 | 127,468 |
| R2 (incl. FE) | 0.043 | 0.585 | 0.606 |
| Within R2 | 0.039 | 0.057 | 0.057 |

```
Note: t-statistics in parentheses
```

## 13.3   Clustering Standard Errors

Apart from biased estimators, we usually have to deal with potentially complex dependencies of our residuals with each other. Such dependencies in the residuals invalidate the i.i.d. assumption of OLS and lead to biased standard errors. With biased OLS standard errors, we cannot reliably interpret the statistical significance of our estimated coefficients.

In our setting, the residuals may be correlated across years for a given firm (time-series dependence), or, alternatively, the residuals may be correlated across different firms (cross-section dependence). One of the most common approaches to dealing with such dependence is the use of *clustered standard errors* (Petersen, 2008). The idea behind clustering is that the correlation of residuals *within* a cluster can be of any form. As the number of clusters grows, the cluster-robust standard errors become consistent (Donald and Lang, 2007; Wooldridge, 2010). A natural requirement for clustering standard errors in practice is hence a sufficiently large number of clusters. Typically, around at least 30 to 50 clusters are seen as sufficient (Cameron et al., 2011).

Instead of relying on the i.i.d. assumption, we can use the `cov_type="clustered"` option in the `fit()`-function as above. The code chunk below applies both one-way clustering by firm as well as two-way clustering by firm and year.

```
model_cluster_firm = lm.PanelOLS.from_formula(
    formula=("investment_lead ~ cash_flows + tobins_q + EntityEffects"
             " + TimeEffects"),
    data=data_investment.set_index(["gvkey", "year"]),
).fit(cov_type="clustered", cluster_entity=True, cluster_time=False)

model_cluster_firmyear = lm.PanelOLS.from_formula(
```

```
formula=("investment_lead ~ cash_flows + tobins_q + EntityEffects"
         " + TimeEffects"),
  data=data_investment.set_index(["gvkey", "year"]),
).fit(cov_type="clustered", cluster_entity=True, cluster_time=True)
```

The table below shows the comparison of the different assumptions behind the standard errors. In the first column, we can see highly significant coefficients on both cash flows and Tobin's q. By clustering the standard errors on the firm level, the $t$-statistics of both coefficients drop in half, indicating a high correlation of residuals within firms. If we additionally cluster by year, we see a drop, particularly for Tobin's q, again. Even after relaxing the assumptions behind our standard errors, both coefficients are still comfortably significant as the $t$-statistics are well above the usual critical values of 1.96 or 2.576 for two-tailed significance tests.

```
prettify_result([
  model_fe_firmyear, model_cluster_firm, model_cluster_firmyear
])
```

| Dependent var. | investment_lead | investment_lead | investment_lead |
|---|---|---|---|
| cash_flows | 0.018 (19.36) | 0.018 (10.57) | 0.018 (9.03) |
| tobins_q | 0.01 (75.42) | 0.01 (33.44) | 0.01 (14.65) |
| | | | |
| Fixed effects | Entity, Time | Entity, Time | Entity, Time |
| VCOV type | Unadjusted | Clustered | Clustered |
| Observations | 127,468 | 127,468 | 127,468 |
| R2 (incl. FE) | 0.606 | 0.606 | 0.606 |
| Within R2 | 0.057 | 0.057 | 0.057 |

Note: t-statistics in parentheses

Inspired by Abadie et al. (2017), we want to close this chapter by highlighting that choosing the right dimensions for clustering is a design problem. Even if the data is informative about whether clustering matters for standard errors, they do not tell you whether you should adjust the standard errors for clustering. Clustering at too aggregate levels can hence lead to unnecessarily inflated standard errors.

## 13.4   Exercises

1. Estimate the two-way fixed effects model with two-way clustered standard errors using quarterly Compustat data from WRDS.
2. Following Peters and Taylor (2017), compute Tobin's q as the market value of outstanding equity mktcap plus the book value of debt (dltt + dlc) minus the current assets atc and everything divided by the book value of property, plant and equipment ppegt. What is the correlation between the measures of Tobin's q? What is the impact on the two-way fixed effects regressions?

# 14

## Difference in Differences

In this chapter, we illustrate the concept of *difference in differences* (DD) estimators by evaluating the effects of climate change regulation on the pricing of bonds across firms. DD estimators are typically used to recover the treatment effects of natural or quasi-natural experiments that trigger sharp changes in the environment of a specific group. Instead of looking at differences in just one group (e.g., the effect in the treated group), DD investigates the treatment effects by looking at the difference between differences in two groups. Such experiments are usually exploited to address endogeneity concerns (e.g., Roberts and Whited, 2013). The identifying assumption is that the outcome variable would change equally in both groups without the treatment. This assumption is also often referred to as the assumption of parallel trends. Moreover, we would ideally also want a random assignment to the treatment and control groups. Due to lobbying or other activities, this randomness is often violated in (financial) economics.

In the context of our setting, we investigate the impact of the Paris Agreement (PA), signed on December 12, 2015, on the bond yields of polluting firms. We first estimate the treatment effect of the agreement using panel regression techniques that we discuss in Chapter 13. We then present two methods to illustrate the treatment effect over time graphically. Although we demonstrate that the treatment effect of the agreement is anticipated by bond market participants well in advance, the techniques we present below can also be applied to many other settings.

The approach we use here replicates the results of Seltzer et al. (2022) partly. Specifically, we borrow their industry definitions for grouping firms into green and brown types. Overall, the literature on environmental, social, and governance (ESG) effects in corporate bond markets is already large but continues to grow (for recent examples, see, e.g., Halling et al. (2021), Handler et al. (2022), Huynh and Xia (2021), among many others).

The current chapter relies on this set of Python packages.

```python
import pandas as pd
import numpy as np
import sqlite3
import linearmodels as lm
import statsmodels.formula.api as smf

from plotnine import *
from scipy.stats import norm
from mizani.breaks import date_breaks
from mizani.formatters import date_format
from regtabletotext import prettify_result
```

Compared to previous chapters, we introduce the `scipy.stats` module from the `scipy` (Virtanen et al., 2020) for simple retrieval of quantiles of the standard normal distribution.

## 14.1   Data Preparation

We use TRACE and Mergent FISD as data sources from our SQLite database introduced in Chapter 3 and Chapter 5.

```python
tidy_finance = sqlite3.connect(database="data/tidy_finance_python.sqlite")

fisd = (pd.read_sql_query(
    sql="SELECT complete_cusip, maturity, offering_amt, sic_code FROM fisd",
    con=tidy_finance,
    parse_dates={"maturity"})
  .dropna()
)

trace_enhanced = (pd.read_sql_query(
    sql=("SELECT cusip_id, trd_exctn_dt, rptd_pr, entrd_vol_qt, yld_pt "
        "FROM trace_enhanced"),
    con=tidy_finance,
    parse_dates={"trd_exctn_dt"})
  .dropna()
)
```

We start our analysis by preparing the sample of bonds. We only consider bonds with a time to maturity of more than one year to the signing of the PA, so that we have sufficient data to analyze the yield behavior after the treatment date. This restriction also excludes all bonds issued after the agreement. We also consider only the first two digits of the SIC industry code to identify the polluting industries (in line with Seltzer et al., 2022).

```python
treatment_date = pd.to_datetime("2015-12-12")
polluting_industries = [
  49, 13, 45, 29, 28, 33, 40, 20, 26, 42, 10, 53, 32, 99, 37
]

bonds = (fisd
  .query("offering_amt > 0 & sic_code != 'None'")
  .assign(
    time_to_maturity=lambda x: (x["maturity"]-treatment_date).dt.days / 365,
    sic_code=lambda x: x["sic_code"].astype(str).str[:2].astype(int),
    log_offering_amt=lambda x: np.log(x["offering_amt"])
  )
  .query("time_to_maturity >= 1")
  .rename(columns={"complete_cusip": "cusip_id"})
  .get(["cusip_id", "time_to_maturity", "log_offering_amt", "sic_code"])
  .assign(polluter=lambda x: x["sic_code"].isin(polluting_industries))
  .reset_index(drop=True)
)
```

Next, we aggregate the individual transactions as reported in TRACE to a monthly panel of bond yields. We consider bond yields for a bond's last trading day in a month. Therefore, we first aggregate bond data to daily frequency and apply common restrictions from the literature (see, e.g., Bessembinder et al., 2008). We weigh each transaction by volume to reflect a trade's relative importance and avoid emphasizing small trades. Moreover, we only consider transactions with reported prices `rptd_pr` larger than 25 (to exclude bonds that are close to default) and only bond-day observations with more than five trades on a corresponding day (to exclude prices based on too few, potentially non-representative transactions).

```python
trace_enhanced = (trace_enhanced
  .query("rptd_pr > 25")
  .assign(weight=lambda x: x["entrd_vol_qt"]*x["rptd_pr"])
  .assign(weighted_yield=lambda x: x["weight"]*x["yld_pt"])
)

trace_aggregated = (trace_enhanced
  .groupby(["cusip_id", "trd_exctn_dt"])
  .aggregate(
    weighted_yield_sum=("weighted_yield", "sum"),
    weight_sum=("weight", "sum"),
    trades=("rptd_pr", "count")
  )
  .reset_index()
  .assign(avg_yield=lambda x: x["weighted_yield_sum"]/x["weight_sum"])
  .dropna(subset=["avg_yield"])
  .query("trades >= 5")
  .assign(trd_exctn_dt=lambda x: pd.to_datetime(x["trd_exctn_dt"]))
  .assign(month=lambda x: x["trd_exctn_dt"]-pd.offsets.MonthBegin())
)

date_index = (trace_aggregated
  .groupby(["cusip_id", "month"])["trd_exctn_dt"]
  .idxmax()
)

trace_aggregated = (trace_aggregated
  .loc[date_index]
  .get(["cusip_id", "month", "avg_yield"])
)
```

By combining the bond-specific information from Mergent FISD for our bond sample with the aggregated TRACE data, we arrive at the main sample for our analysis.

```python
bonds_panel = (bonds
  .merge(trace_aggregated, how="inner", on="cusip_id")
  .dropna()
)
```

Before we can run the first regression, we need to define the `treated` indicator,[1] which is the product of the `post_period` (i.e., all months after the signing of the PA) and the `polluter` indicator defined above.

---

[1]Note that by using a generic name here, everybody can replace ours with their sample data and run the code to produce standard regression tables and illustrations.

```
bonds_panel = (bonds_panel
  .assign(
    post_period=lambda x: (
      x["month"] >= (treatment_date-pd.offsets.MonthBegin())
    )
  )
  .assign(treated=lambda x: x["polluter"] & x["post_period"])
  .assign(month_cat=lambda x: pd.Categorical(x["month"], ordered=True))
)
```

As usual, we tabulate summary statistics of the variables that enter the regression to check the validity of our variable definitions.

```
bonds_panel_summary = (bonds_panel
  .melt(var_name="measure",
        value_vars=["avg_yield", "time_to_maturity", "log_offering_amt"])
  .groupby("measure")
  .describe(percentiles=[0.05, 0.5, 0.95])
)
np.round(bonds_panel_summary, 2)
```

| measure | value count | mean | std | min | 5% | 50% | 95% | max |
|---|---|---|---|---|---|---|---|---|
| avg_yield | 127546.0 | 4.08 | 4.21 | 0.06 | 1.27 | 3.38 | 8.11 | 127.97 |
| log_offering_amt | 127546.0 | 13.27 | 0.82 | 4.64 | 12.21 | 13.22 | 14.51 | 16.52 |
| time_to_maturity | 127546.0 | 8.55 | 8.41 | 1.01 | 1.50 | 5.81 | 27.41 | 100.70 |

## 14.2   Panel Regressions

The PA is a legally binding international treaty on climate change. It was adopted by 196 parties at COP 21 in Paris on December 12, 2015 and entered into force on November 4, 2016. The PA obliges developed countries to support efforts to build clean, climate-resilient futures. One may thus hypothesize that adopting climate-related policies may affect financial markets. To measure the magnitude of this effect, we first run an ordinary least square (OLS) regression without fixed effects where we include the `treated`, `post_period`, and `polluter` dummies, as well as the bond-specific characteristics `log_offering_amt` and `time_to_maturity`. This simple model assumes that there are essentially two periods (before and after the PA) and two groups (polluters and non-polluters). Nonetheless, it should indicate whether polluters have higher yields following the PA compared to non-polluters.

The second model follows the typical DD regression approach by including individual (`cusip_id`) and time (`month`) fixed effects. In this model, we do not include any other variables from the simple model because the fixed effects subsume them, and we observe the coefficient of our main variable of interest: `treated`.

```
model_without_fe = lm.PanelOLS.from_formula(
    formula=("avg_yield ~ treated + post_period + polluter + log_offering_amt"
            " + time_to_maturity + 1"),
    data=bonds_panel.set_index(["cusip_id", "month"]),
).fit()

model_with_fe = lm.PanelOLS.from_formula(
    formula=("avg_yield ~ treated + EntityEffects + TimeEffects"),
    data=bonds_panel.set_index(["cusip_id", "month"]),
).fit()

prettify_result([model_without_fe, model_with_fe])
```

| Dependent var. | avg_yield | avg_yield |
|---|---|---|
| Intercept | 10.733 (57.06) | |
| treated | 0.453 (9.14) | 0.974 (29.3) |
| post_period | -0.178 (-6.04) | |
| polluter | 0.486 (15.43) | |
| log_offering_amt | -0.55 (-38.99) | |
| time_to_maturity | 0.058 (41.53) | |
| | | |
| Fixed effects | | Entity, Time |
| VCOV type | Unadjusted | Unadjusted |
| Observations | 127,546 | 127,546 |
| R2 (incl. FE) | 0.032 | 0.648 |
| Within R2 | 0.004 | 0.012 |

Note: t-statistics in parentheses

Both models indicate that polluters have significantly higher yields after the PA than non-polluting firms. Note that the magnitude of the **treated** coefficient varies considerably across models.

## 14.3 Visualizing Parallel Trends

Even though the regressions above indicate that there is an impact of the PA on bond yields of polluters, the tables do not tell us anything about the dynamics of the treatment effect. In particular, the models provide no indication about whether the crucial *parallel trends* assumption is valid. This assumption requires that in the absence of treatment, the difference between the two groups is constant over time. Although there is no well-defined statistical test for this assumption, visual inspection typically provides a good indication.

To provide such visual evidence, we revisit the simple OLS model and replace the **treated** and **post_period** indicators with month dummies for each group. This approach estimates the average yield change of both groups for each period and provides corresponding confidence intervals. Plotting the coefficient estimates for both groups around the treatment date shows us the dynamics of our panel data.

```python
model_without_fe_time = (smf.ols(
    formula=("avg_yield ~ polluter + month_cat:polluter + time_to_maturity"
             " + log_offering_amt"),
    data=bonds_panel)
  .fit()
  .summary()
)


model_without_fe_coefs = (
  pd.DataFrame(model_without_fe_time.tables[1].data[1:],
              columns=["term", "estimate", "std_error",
                       "t_stat", "p_value", "ci_1", "ci_2"])
  .query("term.str.contains('month_cat')")
  .assign(
    month=lambda x:
      x["term"].str.extract(r"(\d{4}-\d{2}-\d{2} \d{2}:\d{2}:\d{2})")
  )
  .assign(month=lambda x: pd.to_datetime(x["month"]))
  .assign(treatment=lambda x: x["term"].str.contains("True"))
  .assign(estimate=lambda x: x["estimate"].astype(float),
          std_error=lambda x: x["std_error"].astype(float))
  .assign(ci_up=lambda x: x["estimate"]+norm.ppf(0.975)*x["std_error"],
          ci_low=lambda x: x["estimate"]+norm.ppf(0.025)*x["std_error"])
)


polluters_plot = (
  ggplot(model_without_fe_coefs,
         aes(x="month", y="estimate",
             color="treatment", linetype="treatment", shape="treatment")) +
  geom_vline(xintercept=pd.to_datetime(treatment_date) -
             pd.offsets.MonthBegin(), linetype="dashed") +
  geom_hline(yintercept=0, linetype="dashed") +
  geom_errorbar(aes(ymin="ci_low", ymax="ci_up"), alpha=0.5) +
  geom_point() +
  guides(linetype=None) +
  labs(x="", y="Yield", shape="Polluter?", color="Polluter?",
       title="Polluters respond stronger than green firms") +
  scale_linetype_manual(values=["solid", "dashed"]) +
  scale_x_datetime(breaks=date_breaks("1 year"), labels=date_format("%Y"))
)
polluters_plot.draw()
```

Figure 14.1 shows that throughout most of 2014, the yields of the two groups changed in unison. However, starting at the end of 2014, the yields start to diverge, reaching the highest difference around the signing of the PA. Afterward, the yields for both groups fall again, and the polluters arrive at the same level as at the beginning of 2014. The non-polluters, on the other hand, even experience significantly lower yields than polluters after the signing of the agreement.

Instead of plotting both groups using the simple model approach, we can also use the fixed-effects model and focus on the polluter's yield response to the signing relative to the

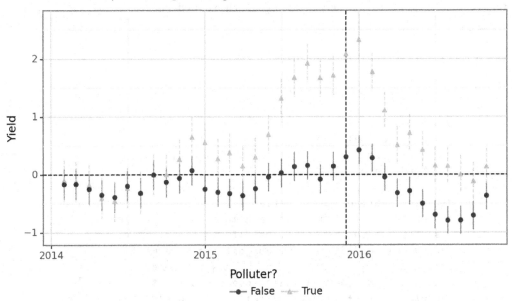

Figure 14.1: The figure shows the coefficient estimates and 95 percent confidence intervals for OLS regressions estimating the treatment effect of the Paris Agreement on bond yields (in percent) for polluters and non-polluters. The horizontal line represents the benchmark yield of polluters before the Paris Agreement. The vertical line indicates the date of the agreement (December 12, 2015).

non-polluters. To perform this estimation, we need to replace the **treated** indicator with separate time dummies for the polluters, each marking a one-month period relative to the treatment date.

```
bonds_panel_alt = (bonds_panel
  .assign(
    diff_to_treatment=lambda x: (
      np.round(
        ((x["month"]-(treatment_date-
            pd.offsets.MonthBegin())).dt.days/365)*12, 0
      ).astype(int)
    )
  )
)

variables = (bonds_panel_alt
  .get(["diff_to_treatment", "month"])
  .drop_duplicates()
  .sort_values("month")
  .copy()
  .assign(variable_name=np.nan)
  .reset_index(drop=True)
)
```

In the next code chunk, we assemble the model formula and regress the monthly yields on the set of time dummies and `cusip_id` and `month` fixed effects.

```python
formula = "avg_yield ~ 1 + "

for j in range(variables.shape[0]):
    if variables["diff_to_treatment"].iloc[j] != 0:
        old_names=list(bonds_panel_alt.columns)

        bonds_panel_alt["new_var"] = (
            bonds_panel_alt["diff_to_treatment"] ==
            variables["diff_to_treatment"].iloc[j]
        ) & bonds_panel_alt["polluter"]

        diff_to_treatment_value=variables["diff_to_treatment"].iloc[j]
        direction="lag" if diff_to_treatment_value < 0 else "lead"
        abs_diff_to_treatment=int(abs(diff_to_treatment_value))
        new_var_name=f"{direction}{abs_diff_to_treatment}"
        variables.at[j, "variable_name"]=new_var_name
        bonds_panel_alt[new_var_name]=bonds_panel_alt["new_var"]
        formula += (f" + {new_var_name}" if j > 0 else new_var_name)

formula = formula + " + EntityEffects + TimeEffects"

model_with_fe_time = (lm.PanelOLS.from_formula(
    formula=formula,
    data=bonds_panel_alt.set_index(["cusip_id", "month"]))
  .fit()
  .summary
)
```

We then collect the regression results into a dataframe that contains the estimates and corresponding 95 percent confidence intervals. Note that we also add a row with zeros for the (omitted) reference point of the time dummies.

```python
lag0_row = pd.DataFrame({
  "term": ["lag0"],
  "estimate": [0],
  "ci_1": [0],
  "ci_2": [0],
  "ci_up": [0],
  "ci_low": [0],
  "month": [treatment_date - pd.offsets.MonthBegin()]
})

model_with_fe_time_coefs = (
  pd.DataFrame(model_with_fe_time.tables[1].data[1:],
              columns=["term", "estimate", "std_error",
                       "t_stat", "p_value", "ci_1", "ci_2"])
  .assign(term=lambda x: x["term"].str.replace("[T.True]", ""))
  .assign(estimate=lambda x: x["estimate"].astype(float),
          std_error=lambda x: x["std_error"].astype(float))
```

```
 .assign(ci_up=lambda x: x["estimate"] + norm.ppf(0.975)*x["std_error"],
          ci_low=lambda x: x["estimate"] + norm.ppf(0.025)*x["std_error"])
 .merge(variables, how="left", left_on="term", right_on="variable_name")
 .drop(columns="variable_name")
 .query("term != 'Intercept'")
)

model_with_fe_time_coefs = pd.concat(
  [model_with_fe_time_coefs, lag0_row],
  ignore_index=True
)
```

Figure 15.2 shows the resulting coefficient estimates.

```
polluter_plot = (
  ggplot(model_with_fe_time_coefs, aes(x="month", y="estimate")) +
  geom_vline(aes(xintercept=treatment_date - pd.offsets.MonthBegin()),
              linetype="dashed") +
  geom_hline(aes(yintercept=0), linetype="dashed") +
  geom_errorbar(aes(ymin="ci_low", ymax="ci_up"), alpha=0.5) +
  geom_point(aes(y="estimate")) +
  labs(x="", y="Yield",
       title="Polluters' yield patterns around Paris Agreement signing") +
  scale_x_datetime(breaks=date_breaks("1 year"), labels=date_format("%Y"))
)
polluter_plot.draw()
```

The resulting graph shown in Figure 15.2 confirms the main conclusion of the previous image: polluters' yield patterns show a considerable anticipation effect starting toward the end of 2014. Yields only marginally increase after the signing of the agreement. However, as opposed to the simple model, we do not see a complete reversal back to the pre-agreement level. Yields of polluters stay at a significantly higher level even one year after the signing.

Notice that during the year after the PA was signed, the 45th president of the United States was elected (on November 8, 2016). During his campaign there were some indications of intentions to withdraw the US from the PA, which ultimately happened on November 4, 2020. Hence, reversal effects are potentially driven by these actions.

## 14.4 Exercises

1. The 46th president of the US rejoined the Paris Agreement on February 19, 2021. Repeat the difference in differences analysis for the day of his election victory. Note that you will also have to download new TRACE data. How did polluters' yields react to this action?

2. Based on the exercise on ratings in Chapter 5, include ratings as a control variable in the analysis above. Do the results change?

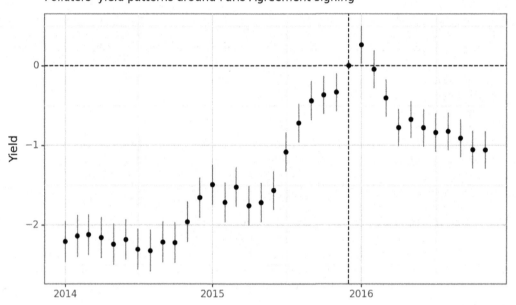

Figure 14.2: The figure shows the coefficient estimates and 95 percent confidence intervals for OLS regressions estimating the treatment effect of the Paris Agreement on bond yields (in percent) for polluters. The horizontal line represents the benchmark yield of polluters before the Paris Agreement. The vertical line indicates the date of the agreement (December 12, 2015).

# 15

## *Factor Selection via Machine Learning*

The aim of this chapter is twofold. From a data science perspective, we introduce `scikit-learn`, a collection of packages for modeling and machine learning (ML). `scikit-learn` comes with a handy workflow for all sorts of typical prediction tasks. From a finance perspective, we address the notion of *factor zoo* (Cochrane, 2011) using ML methods. We introduce Lasso, Ridge, and Elastic Net regression as a special case of penalized regression models. Then, we explain the concept of cross-validation for model *tuning* with Elastic Net regularization as a popular example. We implement and showcase the entire cycle from model specification, training, and forecast evaluation within the `scikit-learn` universe. While the tools can generally be applied to an abundance of interesting asset pricing problems, we apply penalized regressions for identifying macroeconomic variables and asset pricing factors that help explain a cross-section of industry portfolios.

In previous chapters, we illustrate that stock characteristics such as size provide valuable pricing information in addition to the market beta. Such findings question the usefulness of the Capital Asset Pricing Model. In fact, during the last decades, financial economists discovered a plethora of additional factors which may be correlated with the marginal utility of consumption (and would thus deserve a prominent role in pricing applications). The search for factors that explain the cross-section of expected stock returns has produced hundreds of potential candidates, as noted more recently by Harvey et al. (2016), Harvey (2017), Mclean and Pontiff (2016), and Hou et al. (2020). Therefore, given the multitude of proposed risk factors, the challenge these days rather is: *do we believe in the relevance of hundreds of risk factors?* During recent years, promising methods from the field of ML got applied to common finance applications. We refer to Mullainathan and Spiess (2017) for a treatment of ML from the perspective of an econometrician, Nagel (2021) for an excellent review of ML practices in asset pricing, Easley et al. (2020) for ML applications in (high-frequency) market microstructure, and Dixon et al. (2020) for a detailed treatment of all methodological aspects.

## 15.1 Brief Theoretical Background

This is a book about *doing* empirical work in a tidy manner, and we refer to any of the many excellent textbook treatments of ML methods and especially penalized regressions for some deeper discussion. Excellent material is provided, for instance, by Hastie et al. (2009), Gareth et al. (2013), and De Prado (2018). Instead, we briefly summarize the idea of Lasso and Ridge regressions as well as the more general Elastic Net. Then, we turn to the fascinating question on *how* to implement, tune, and use such models with the `scikit-learn` package.

To set the stage, we start with the definition of a linear model: Suppose we have data $(y_t, x_t), t = 1, \ldots, T$, where $x_t$ is a $(K \times 1)$ vector of regressors and $y_t$ is the response for observation $t$. The linear model takes the form $y_t = \beta' x_t + \varepsilon_t$ with some error term $\varepsilon_t$ and has been studied in abundance. For $K \leq T$, the well-known ordinary-least square (OLS) estimator for the $(K \times 1)$ vector $\beta$ minimizes the sum of squared residuals and is then

$$\hat{\beta}^{\text{ols}} = \left( \sum_{t=1}^{T} x_t' x_t \right)^{-1} \sum_{t=1}^{T} x_t' y_t. \tag{15.1}$$

While we are often interested in the estimated coefficient vector $\hat{\beta}^{\text{ols}}$, ML is about the predictive performance most of the time. For a new observation $\tilde{x}_t$, the linear model generates predictions such that

$$\hat{y}_t = E\left(y | x_t = \tilde{x}_t\right) = \hat{\beta}^{\text{ols}'} \tilde{x}_t. \tag{15.2}$$

Is this the best we can do? Not necessarily: instead of minimizing the sum of squared residuals, penalized linear models can improve predictive performance by choosing other estimators $\hat{\beta}$ with lower variance than the estimator $\hat{\beta}^{\text{ols}}$. At the same time, it seems appealing to restrict the set of regressors to a few meaningful ones, if possible. In other words, if $K$ is large (such as for the number of proposed factors in the asset pricing literature), it may be a desirable feature to *select* reasonable factors and set $\hat{\beta}_k^{\text{ols}} = 0$ for some redundant factors.

It should be clear that the promised benefits of penalized regressions, i.e., reducing the mean squared error (MSE), come at a cost. In most cases, reducing the variance of the estimator introduces a bias such that $E\left(\hat{\beta}\right) \neq \beta$. What is the effect of such a bias-variance trade-off? To understand the implications, assume the following data-generating process for $y$:

$$y = f(x) + \varepsilon, \quad \varepsilon \sim \left(0, \sigma_\varepsilon^2\right) \tag{15.3}$$

We want to recover $f(x)$, which denotes some unknown functional which maps the relationship between $x$ and $y$. While the properties of $\hat{\beta}^{\text{ols}}$ as an unbiased estimator may be desirable under some circumstances, they are certainly not if we consider predictive accuracy. Alternative predictors $\hat{f}(x)$ could be more desirable: For instance, the MSE depends on our model choice as follows:

$$
\begin{aligned}
MSE &= E\left(\left(y - \hat{f}(x)\right)^2\right) = E\left(\left(f(x) + \epsilon - \hat{f}(x)\right)^2\right) \\
&= \underbrace{E\left(\left(f(x) - \hat{f}(x)\right)^2\right)}_{\text{total quadratic error}} + \underbrace{E\left(\epsilon^2\right)}_{\text{irreducible error}} \\
&= E\left(\hat{f}(x)^2\right) + E\left(f(x)^2\right) - 2E\left(f(x)\hat{f}(x)\right) + \sigma_\varepsilon^2 \\
&= E\left(\hat{f}(x)^2\right) + f(x)^2 - 2f(x)E\left(\hat{f}(x)\right) + \sigma_\varepsilon^2 \\
&= \underbrace{\text{Var}\left(\hat{f}(x)\right)}_{\text{variance of model}} + \underbrace{E\left(\left(f(x) - \hat{f}(x)\right)\right)^2}_{\text{squared bias}} + \sigma_\varepsilon^2.
\end{aligned}
\tag{15.4}
$$

While no model can reduce $\sigma_\varepsilon^2$, a biased estimator with small variance may have a lower MSE than an unbiased estimator.

## 15.1.1 Ridge regression

One biased estimator is known as Ridge regression. Hoerl and Kennard (1970) propose to minimize the sum of squared errors *while simultaneously imposing a penalty on the $L_2$ norm of the parameters $\hat{\beta}$*. Formally, this means that for a penalty factor $\lambda \geq 0$, the minimization problem takes the form $\min_\beta (y - X\beta)' (y - X\beta)$ s.t. $\beta'\beta \leq c$. Here $c \geq 0$ is a constant that depends on the choice of $\lambda$. The larger $\lambda$, the smaller $c$ (technically speaking, there is a one-to-one relationship between $\lambda$, which corresponds to the Lagrangian of the minimization problem above and $c$). Here, $X = (x_1 \dots x_T)'$ and $y = (y_1, \dots, y_T)'$. A closed-form solution for the resulting regression coefficient vector $\beta^{\text{ridge}}$ exists:

$$\hat{\beta}^{\text{ridge}} = (X'X + \lambda I)^{-1} X'y, \tag{15.5}$$

where $I$ is the identity matrix of dimension $K$. A couple of observations are worth noting: $\hat{\beta}^{\text{ridge}} = \hat{\beta}^{\text{ols}}$ for $\lambda = 0$ and $\hat{\beta}^{\text{ridge}} \to 0$ for $\lambda \to \infty$. Also for $\lambda > 0$, $(X'X + \lambda I)$ is non-singular even if $X'X$ is which means that $\hat{\beta}^{\text{ridge}}$ exists even if $\hat{\beta}$ is not defined. However, note also that the Ridge estimator requires careful choice of the hyperparameter $\lambda$ which controls the *amount of regularization*: a larger value of $\lambda$ implies *shrinkage* of the regression coefficient toward 0; a smaller value of $\lambda$ reduces the bias of the resulting estimator.

Note that $X$ usually contains an intercept column with ones. As a general rule, the associated intercept coefficient is not penalized. In practice, this often implies that $y$ is simply demeaned before computing $\hat{\beta}^{\text{ridge}}$.

What about the statistical properties of the Ridge estimator? First, the bad news is that $\hat{\beta}^{\text{ridge}}$ is a biased estimator of $\beta$. However, the good news is that (under homoscedastic error terms) the variance of the Ridge estimator is guaranteed to be *smaller* than the variance of the OLS estimator. We encourage you to verify these two statements in the Exercises. As a result, we face a trade-off: The Ridge regression sacrifices some unbiasedness to achieve a smaller variance than the OLS estimator.

## 15.1.2 Lasso

An alternative to Ridge regression is the Lasso (*least absolute shrinkage and selection operator*). Similar to Ridge regression, the Lasso (Tibshirani, 1996) is a penalized and biased estimator. The main difference to Ridge regression is that Lasso does not only *shrink* coefficients but effectively selects variables by setting coefficients for *irrelevant* variables to zero. Lasso implements a $L_1$ penalization on the parameters such that:

$$\hat{\beta}^{\text{Lasso}} = \arg\min_\beta (Y - X\beta)' (Y - X\beta) \text{ s.t. } \sum_{k=1}^{K} |\beta_k| < c(\lambda). \tag{15.6}$$

There is no closed-form solution for $\hat{\beta}^{\text{Lasso}}$ in the above maximization problem, but efficient algorithms exist (e.g., the `glmnet` package for R and Python). Like for Ridge regression, the hyperparameter $\lambda$ has to be specified beforehand.

The corresponding Lagrangian reads as follows

$$\hat{\beta}_\lambda^{\text{Lasso}} = \arg\min_\beta (Y - X\beta)' (Y - X\beta) + \lambda \sum_{k=1}^{K} |\beta_k|. \tag{15.7}$$

### 15.1.3   Elastic Net

The Elastic Net (Zou and Hastie, 2005) combines $L_1$ with $L_2$ penalization and encourages a grouping effect, where strongly correlated predictors tend to be in or out of the model together. In terms of the Lagrangian, this more general framework considers the following optimization problem:

$$\hat{\beta}^{\text{EN}} = \arg\min_{\beta} (Y - X\beta)' (Y - X\beta) + \lambda(1 - \rho) \sum_{k=1}^{K} |\beta_k| + \frac{1}{2}\lambda\rho \sum_{k=1}^{K} \beta_k^2 \qquad (15.8)$$

Now, we have to choose two hyperparameters: the *shrinkage* factor $\lambda$ and the *weighting parameter* $\rho$. The Elastic Net resembles Lasso for $\rho = 0$ and Ridge regression for $\rho = 1$. While the `glmnet` package provides efficient algorithms to compute the coefficients of penalized regressions, it is a good exercise to implement Ridge and Lasso estimation on your own before you use the `scikit-learn` back-end.

## 15.2   Python Packages

To get started, we load the required packages and data. The main focus is on the workflow behind the `scikit-learn` (Pedregosa et al., 2011) package collection.

```
import pandas as pd
import numpy as np
import sqlite3

from plotnine import *
from mizani.formatters import percent_format, date_format
from mizani.breaks import date_breaks
from itertools import product
from sklearn.model_selection import (
  train_test_split, GridSearchCV, TimeSeriesSplit, cross_val_score
)
from sklearn.compose import ColumnTransformer
from sklearn.preprocessing import StandardScaler
from sklearn.pipeline import Pipeline
from sklearn.linear_model import ElasticNet, Lasso, Ridge
```

## 15.3   Data Preparation

In this analysis, we use four different data sources that we load from our SQLite database introduced in Chapter 3. We start with two different sets of factor portfolio returns which have been suggested as representing practical risk factor exposure and thus should be relevant when it comes to asset pricing applications.

- The standard workhorse: monthly Fama-French 3 factor returns (market, small-minus-big, and high-minus-low book-to-market valuation sorts) defined in Fama and French (1992) and Fama and French (1993).

- Monthly q-factor returns from Hou et al. (2014). The factors contain the size factor, the investment factor, the return-on-equity factor, and the expected growth factor.

Next, we include macroeconomic predictors which may predict the general stock market economy. Macroeconomic variables effectively serve as conditioning information such that their inclusion hints at the relevance of conditional models instead of unconditional asset pricing. We refer the interested reader to Cochrane (2009) on the role of conditioning information.

- Our set of macroeconomic predictors comes from Welch and Goyal (2008). The data has been updated by the authors until 2021 and contains monthly variables that have been suggested as good predictors for the equity premium. Some of the variables are the dividend price ratio, earnings price ratio, stock variance, net equity expansion, treasury bill rate, and inflation.

Finally, we need a set of *test assets*. The aim is to understand which of the plenty factors and macroeconomic variable combinations prove helpful in explaining our test assets' cross-section of returns. In line with many existing papers, we use monthly portfolio returns from ten different industries according to the definition from Kenneth French's homepage[1] as test assets.

```python
tidy_finance = sqlite3.connect(database="data/tidy_finance_python.sqlite")

factors_ff3_monthly = (pd.read_sql_query(
    sql="SELECT * FROM factors_ff3_monthly",
    con=tidy_finance,
    parse_dates={"month"})
  .add_prefix("factor_ff_")
)

factors_q_monthly = (pd.read_sql_query(
    sql="SELECT * FROM factors_q_monthly",
    con=tidy_finance,
    parse_dates={"month"})
  .add_prefix("factor_q_")
)

macro_predictors = (pd.read_sql_query(
    sql="SELECT * FROM macro_predictors",
    con=tidy_finance,
    parse_dates={"month"})
 .add_prefix("macro_")
)

industries_ff_monthly = (pd.read_sql_query(
    sql="SELECT * FROM industries_ff_monthly",
    con=tidy_finance,
    parse_dates={"month"})
  .melt(id_vars="month", var_name="industry", value_name="ret")
)
```

---

[1] https://mba.tuck.dartmouth.edu/pages/faculty/ken.french/Data_Library/det_10_ind_port.html

We combine all the monthly observations into one dataframe.

```
data = (industries_ff_monthly
  .merge(factors_ff3_monthly,
         how="left", left_on="month", right_on="factor_ff_month")
  .merge(factors_q_monthly,
         how="left", left_on="month", right_on="factor_q_month")
  .merge(macro_predictors,
         how="left", left_on="month", right_on="macro_month")
  .assign(ret_excess=lambda x: x["ret"] - x["factor_ff_rf"])
  .drop(columns=["ret", "factor_ff_month", "factor_q_month", "macro_month"])
  .dropna()
)
```

Our data contains 22 columns of regressors with the 13 macro-variables and 8 factor returns for each month. Figure 15.1 provides summary statistics for the 10 monthly industry excess returns in percent. One can see that the dispersion in the excess returns varies widely across industries.

```
data_plot = (ggplot(data,
  aes(x="industry", y="ret_excess")) +
  geom_boxplot() +
  coord_flip() +
  labs(x="", y="",
       title="Excess return distributions by industry in percent") +
    scale_y_continuous(labels=percent_format())
)
data_plot.draw()
```

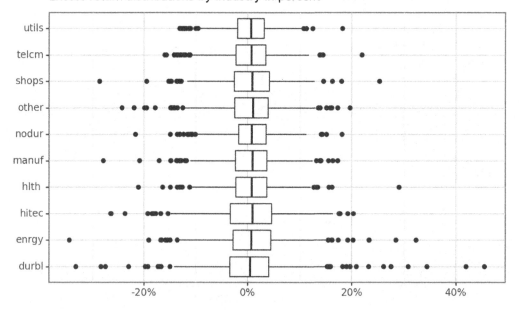

Figure 15.1: The box plots show the monthly dispersion of returns for 10 different industries.

## 15.4   Machine Learning Workflow

To illustrate penalized linear regressions, we employ the `scikit-learn` collection of packages for modeling and ML. Using the ideas of Ridge and Lasso regressions, the following example guides you through (i) pre-processing the data (data split and variable mutation), (ii) building models, (iii) fitting models, and (iv) tuning models to create the "best" possible predictions.

### 15.4.1   Pre-process data

We want to explain excess returns with all available predictors. The regression equation thus takes the form

$$r_t = \alpha_0 + \left( \tilde{f}_t \otimes \tilde{z}_t \right) B + \varepsilon_t \tag{15.9}$$

where $r_t$ is the vector of industry excess returns at time $t$, $\otimes$ denotes the Kronecker product and $\tilde{f}_t$ and $\tilde{z}_t$ are the (standardized) vectors of factor portfolio returns and macroeconomic variables.

We hence perform the following pre-processing steps:

- We exclude the column *month* from the analysis
- We include all interaction terms between factors and macroeconomic predictors
- We demean and scale each regressor such that the standard deviation is one

Scaling is often necessary in machine learning applications, especially when combining variables of different magnitudes or units, or when using algorithms sensitive to feature scales (e.g., gradient descent-based algorithms). We use `ColumnTransformer()` to scale all regressors using `StandardScaler()`. The `remainder="drop"` ensures that only the specified columns are retained in the output, and others are dropped. The option `verbose_feature_names_out=False` ensures that the output feature names remain unchanged. Also note that we use the `zip()` function to pair each element from `column_names` with its corresponding list of values from `new_column_values`, creating tuples, and then convert these tuples into a dictionary using `dict()` from which we create a dataframe.

```python
macro_variables = data.filter(like="macro").columns
factor_variables = data.filter(like="factor").columns

column_combinations = list(product(macro_variables, factor_variables))

new_column_values = []
for macro_column, factor_column in column_combinations:
    new_column_values.append(data[macro_column] * data[factor_column])

column_names = [" x ".join(t) for t in column_combinations]
new_columns = pd.DataFrame(dict(zip(column_names, new_column_values)))

data = pd.concat([data, new_columns], axis=1)

preprocessor = ColumnTransformer(
  transformers=[
    ("scale", StandardScaler(),
```

```
    [col for col in data.columns
        if col not in ["ret_excess", "month", "industry"]])
    ],
    remainder="drop",
    verbose_feature_names_out=False
)
```

## 15.4.2  Build a model

Next, we can build an actual model based on our pre-processed data. In line with the definition above, we estimate regression coefficients of a Lasso regression such that we get

$$\hat{\beta}_\lambda^{\text{Lasso}} = \arg \min_\beta \left(Y - X\beta\right)' \left(Y - X\beta\right) + \lambda \sum_{k=1}^{K} |\beta_k|. \qquad (15.10)$$

In the application at hand, $X$ contains 104 columns with all possible interactions between factor returns and macroeconomic variables. We want to emphasize that the workflow for *any* model is very similar, irrespective of the specific model. As you will see further below, it is straightforward to fit Ridge regression coefficients and, later, Neural networks or Random forests with similar code. For now, we start with the linear regression model with an arbitrarily chosen value for the penalty factor $\lambda$ (denoted as `alpha=0.007` in the code below). In the setup below, `l1_ratio` denotes the value of $1 - \rho$, hence setting `l1_ratio=1` implies the Lasso.

```
lm_model = ElasticNet(
    alpha=0.007,
    l1_ratio=1,
    max_iter=5000,
    fit_intercept=False
)

lm_pipeline = Pipeline([
    ("preprocessor", preprocessor),
    ("regressor", lm_model)
])
```

That's it - we are done! The object `lm_model_pipeline` contains the definition of our model with all required information, in particular the pre-processing steps and the regression model.

## 15.4.3  Fit a model

With the pipeline from above, we are ready to fit it to the data. Typically, we use training data to fit the model. The training data is pre-processed according to our recipe steps, and the Lasso regression coefficients are computed. For illustrative purposes, we focus on the manufacturing industry for now.

```
data_manufacturing = data.query("industry == 'manuf'")
training_date = "2011-12-01"

data_manufacturing_training = (data_manufacturing
```

```
  .query(f"month<'{training_date}'")
)

lm_fit = lm_pipeline.fit(
  data_manufacturing_training,
  data_manufacturing_training.get("ret_excess")
)
```

First, we focus on the in-sample predicted values $\hat{y}_t = x_t \hat{\beta}^{\text{Lasso}}$. Figure 15.2 illustrates the projections for the *entire* time series of the manufacturing industry portfolio returns.

```
predicted_values = (pd.DataFrame({
    "Fitted value": lm_fit.predict(data_manufacturing),
    "Realization": data_manufacturing.get("ret_excess")
  })
  .assign(month = data_manufacturing["month"])
  .melt(id_vars="month", var_name="Variable", value_name="return")
)

predicted_values_plot = (
  ggplot(predicted_values,
        aes(x="month", y="return",
            color="Variable", linetype="Variable")) +
  annotate(
    "rect",
    xmin=data_manufacturing_training["month"].max(),
    xmax=data_manufacturing["month"].max(),
    ymin=-np.inf, ymax=np.inf,
    alpha=0.25, fill="#808080"
  ) +
  geom_line() +
  labs(x="", y="", color="", linetype="",
       title="Monthly realized and fitted manufacturing risk premia") +
  scale_x_datetime(breaks=date_breaks("5 years"),
                   labels=date_format("%Y")) +
  scale_y_continuous(labels=percent_format())
)
predicted_values_plot.draw()
```

What do the estimated coefficients look like? To analyze these values, it is worth computing the coefficients $\hat{\beta}^{\text{Lasso}}$ directly. The code below estimates the coefficients for the Lasso and Ridge regression for the processed training data sample for a grid of different $\lambda$'s.

```
x = preprocessor.fit_transform(data_manufacturing)
y = data_manufacturing["ret_excess"]

alphas = np.logspace(-5, 5, 100)

coefficients_lasso = []
for a in alphas:
    lasso = Lasso(alpha=a, fit_intercept=False)
    coefficients_lasso.append(lasso.fit(x, y).coef_)
```

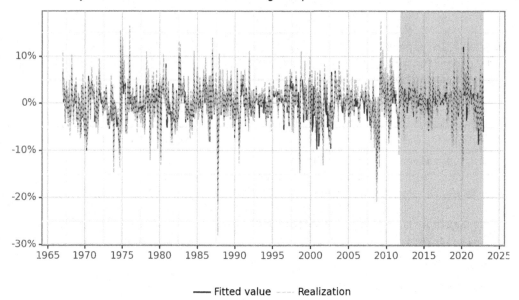

Figure 15.2: The figure shows monthly realized and fitted manufacturing industry risk premium. The grey area corresponds to the out of sample period.

```
coefficients_lasso = (pd.DataFrame(coefficients_lasso)
  .assign(alpha=alphas, model="Lasso")
  .melt(id_vars=["alpha", "model"])
)

coefficients_ridge = []
for a in alphas:
    ridge = Ridge(alpha=a, fit_intercept=False)
    coefficients_ridge.append(ridge.fit(x, y).coef_)

coefficients_ridge = (pd.DataFrame(coefficients_ridge)
  .assign(alpha=alphas, model="Ridge")
  .melt(id_vars=["alpha", "model"])
)
```

The dataframes `lasso_coefficients` and `ridge_coefficients` contain an entire sequence of estimated coefficients for multiple values of the penalty factor $\lambda$. Figure 15.3 illustrates the trajectories of the regression coefficients as a function of the penalty factor. Both Lasso and Ridge coefficients converge to zero as the penalty factor increases.

```
coefficients_plot = (
  ggplot(pd.concat([coefficients_lasso, coefficients_ridge]),
         aes(x="alpha", y="value", color="variable")) +
  geom_line() +
  facet_wrap("model") +
```

```
    labs(x="Penalty factor (lambda)", y="",
         title="Estimated coefficient paths for different penalty factors") +
    scale_x_log10() +
    theme(legend_position="none"))
coefficients_plot.draw()
```

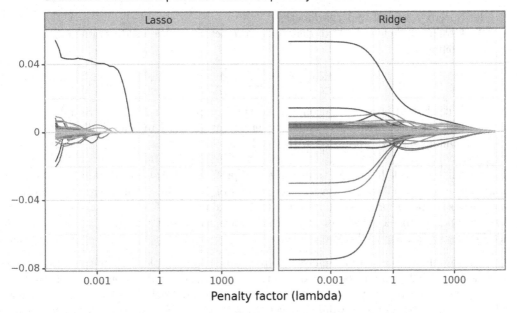

Figure 15.3: The figure shows estimated coefficient paths for different penalty factors. The penalty parameters are chosen iteratively to resemble the path from no penalization to a model that excludes all variables.

### 15.4.4 Tune a model

To compute $\hat{\beta}_\lambda^{\text{Lasso}}$ , we simply imposed an arbitrary value for the penalty hyperparameter $\lambda$. Model tuning is the process of optimally selecting such hyperparameters through *cross-validation*.

The goal for choosing $\lambda$ (or any other hyperparameter, e.g., $\rho$ for the Elastic Net) is to find a way to produce predictors $\hat{Y}$ for an outcome $Y$ that minimizes the mean squared prediction error $\text{MSPE} = E\left(\frac{1}{T}\sum_{t=1}^{T}(\hat{y}_t - y_t)^2\right)$. Unfortunately, the MSPE is not directly observable. We can only compute an estimate because our data is random and because we do not observe the entire population.

Obviously, if we train an algorithm on the same data that we use to compute the error, our estimate MSPE would indicate way better predictive accuracy than what we can expect in real out-of-sample data. The result is called overfitting.

Cross-validation is a technique that allows us to alleviate this problem. We approximate the true MSPE as the average of many MSPE obtained by creating predictions for $K$ new random samples of the data, none of them used to train the algorithm

$\frac{1}{K}\sum_{k=1}^{K}\frac{1}{T}\sum_{t=1}^{T}\left(\hat{y}_t^k - y_t^k\right)^2$. In practice, this is done by carving out a piece of our data and pretending it is an independent sample. We again divide the data into a training set and a test set. The MSPE on the test set is our measure for actual predictive ability, while we use the training set to fit models with the aim to find the *optimal* hyperparameter values. To do so, we further divide our training sample into (several) subsets, fit our model for a grid of potential hyperparameter values (e.g., $\lambda$), and evaluate the predictive accuracy on an *independent* sample. This works as follows:

1. Specify a grid of hyperparameters
2. Obtain predictors $\hat{y}_i(\lambda)$ to denote the predictors for the used parameters $\lambda$
3. Compute

$$\text{MSPE}(\lambda) = \frac{1}{K}\sum_{k=1}^{K}\frac{1}{T}\sum_{t=1}^{T}\left(\hat{y}_t^k(\lambda) - y_t^k\right)^2. \tag{15.11}$$

With K-fold cross-validation, we do this computation $K$ times. Simply pick a validation set with $M = T/K$ observations at random and think of these as random samples $y_1^k, \ldots, y_{\frac{T}{K}}^k$, with $k = 1$.

How should you pick $K$? Large values of $K$ are preferable because the training data better imitates the original data. However, larger values of $K$ will have much higher computation time. `scikit-learn` provides all required tools to conduct $K$-fold cross-validation. We just have to update our model specification. In our case, we specify the penalty factor $\lambda$ as well as the mixing factor $\rho$ as *free* parameters.

For our sample, we consider a time-series cross-validation sample. This means that we tune our models with 20 random samples of length five years with a validation period of four years. For a grid of possible hyperparameters, we then fit the model for each fold and evaluate MSPE in the corresponding validation set. Finally, we select the model specification with the lowest MSPE in the validation set. First, we define the cross-validation folds based on our training data only.

Then, we evaluate the performance for a grid of different penalty values. `scikit-learn` provides functionalities to construct a suitable grid of hyperparameters with `GridSearchCV()`. The code chunk below creates a $10 \times 3$ hyperparameters grid. Then, the method `fit()` evaluates all the models for each fold.

```
initial_years = 5
assessment_months = 48
n_splits = int(len(data_manufacturing)/assessment_months) - 1
length_of_year = 12
alphas = np.logspace(-6, 2, 100)

data_folds = TimeSeriesSplit(
  n_splits=n_splits,
  test_size=assessment_months,
  max_train_size=initial_years * length_of_year
)

params = {
  "regressor__alpha": alphas,
  "regressor__l1_ratio": (0.0, 0.5, 1)
}
```

```
finder = GridSearchCV(
  lm_pipeline,
  param_grid=params,
  scoring="neg_root_mean_squared_error",
  cv=data_folds
)

finder = finder.fit(
  data_manufacturing, data_manufacturing.get("ret_excess")
)
```

After the tuning process, we collect the evaluation metrics (the root mean-squared error in our example) to identify the *optimal* model. Figure 15.4 illustrates the average validation set's root mean-squared error for each value of $\lambda$ and $\rho$.

```
validation = (pd.DataFrame(finder.cv_results_)
  .assign(
    mspe=lambda x: -x["mean_test_score"],
    param_regressor__alpha=lambda x: pd.to_numeric(
      x["param_regressor__alpha"], errors="coerce"
    )
  )
)

validation_plot = (ggplot(validation,
  aes(x="param_regressor__alpha", y="mspe",
      color="param_regressor__l1_ratio",
      shape="param_regressor__l1_ratio",
      group="param_regressor__l1_ratio")) +
  geom_point() +
  geom_line() +
  labs(x ="Penalty factor (lambda)", y="Root MSPE",
       title="Root MSPE for different penalty factors",
       color="Proportion of Lasso Penalty",
       shape="Proportion of Lasso Penalty") +
  scale_x_log10() +
  guides(linetype="none")
)
validation_plot.draw()
```

Figure 15.4 shows that the MSPE drops faster for Lasso and Elastic Net compared to Ridge regressions as penalty factor increases. However, for higher penalty factors, the MSPE for Ridge regressions dips below the others, which both slightly increase again above a certain threshold. Recall that the larger the regularization, the more restricted the model becomes. The best performing model yields a penalty parameter (`alpha`) of 0.0063 and a mixture factor ($\rho$) of 0.5.

## 15.4.5 Full workflow

Our starting point was the question: Which factors determine industry returns? While Avramov et al. (2023) provide a Bayesian analysis related to the research question above,

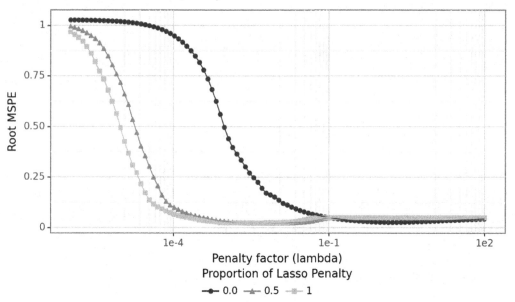

Figure 15.4: The figure shows root MSPE for different penalty factors. Evaluation of manufacturing excess returns for different penalty factors (lambda) and proportions of Lasso penalty (rho). 1.0 indicates Lasso, 0.5 indicates Elastic Net, and 0.0 indicates Ridge.

we choose a simplified approach: To illustrate the entire workflow, we now run the penalized regressions for all ten industries. We want to identify relevant variables by fitting Lasso models for each industry returns time series. More specifically, we perform cross-validation for each industry to identify the optimal penalty factor $\lambda$.

First, we define the Lasso model with one tuning parameter.

```
lm_model = Lasso(fit_intercept=False, max_iter=5000)

params = {"regressor__alpha": alphas}

lm_pipeline = Pipeline([
  ("preprocessor", preprocessor),
  ("regressor", lm_model)
])
```

The following task can be easily parallelized to reduce computing time, but we use a simple loop for ease of exposition.

```
all_industries = data["industry"].drop_duplicates()

results = []
for industry in all_industries:
  print(industry)
  finder = GridSearchCV(
    lm_pipeline,
```

```
    param_grid=params,
    scoring="neg_mean_squared_error",
    cv=data_folds
  )

  finder = finder.fit(
    data.query("industry == @industry"),
    data.query("industry == @industry").get("ret_excess")
  )
  results.append(
    pd.DataFrame(finder.best_estimator_.named_steps.regressor.coef_ != 0)
  )

selected_factors = (
  pd.DataFrame(
    lm_pipeline[:-1].get_feature_names_out(),
    columns=["variable"]
  )
  .assign(variable = lambda x: (
    x["variable"].str.replace("factor_|ff_|q_|macro_","")))
  )
  .assign(**dict(zip(all_industries, results)))
  .melt(id_vars="variable", var_name ="industry")
  .query("value == True")
)
```

What has just happened? In principle, exactly the same as before but instead of computing the Lasso coefficients for one industry, we did it for ten sequentially. Now, we just have to do some housekeeping and keep only variables that Lasso does *not* set to zero. We illustrate the results in a heat map in Figure 15.5.

```
selected_factors_plot = (
  ggplot(selected_factors,
         aes(x="variable", y="industry")) +
  geom_tile() +
  labs(x="", y="",
       title="Selected variables for different industries") +
  coord_flip() +
  scale_x_discrete(limits=reversed) +
  theme(axis_text_x=element_text(rotation=70, hjust=1),
        figure_size=(6.4, 6.4))
)
selected_factors_plot.draw()
```

The heat map in Figure 15.5 conveys two main insights: first, we see that many factors, macroeconomic variables, and interaction terms are not relevant for explaining the cross-section of returns across the industry portfolios. In fact, only `factor_ff_mkt_excess` and its interaction with `macro_bm` a role for several industries. Second, there seems to be quite some heterogeneity across different industries. While barely any variable is selected by Lasso for Utilities, many factors are selected for, e.g., Durable and Manufacturing, but the selected factors do not necessarily coincide. In other words, there seems to be a clear picture that we do not need many factors, but Lasso does not provide a factor that consistently provides pricing abilities across industries.

## 15.5  Exercises

1. Write a function that requires three inputs, namely, y (a $T$ vector), X (a $(T \times K)$ matrix), and `lambda` and then returns the Ridge estimator (a $K$ vector) for a given penalization parameter $\lambda$. Recall that the intercept should not be penalized. Therefore, your function should indicate whether $X$ contains a vector of ones as the first column, which should be exempt from the $L_2$ penalty.

2. Compute the $L_2$ norm ($\beta'\beta$) for the regression coefficients based on the predictive regression from the previous exercise for a range of $\lambda$'s and illustrate the effect of penalization in a suitable figure.

3. Now, write a function that requires three inputs, namely, y (a $T$ vector), X (a $(T \times K)$ matrix), and $\lambda$ and then returns the Lasso estimator (a $K$ vector) for a given penalization parameter $\lambda$. Recall that the intercept should not be penalized. Therefore, your function should indicate whether $X$ contains a vector of ones as the first column, which should be exempt from the $L_1$ penalty.

4. After you understand what Ridge and Lasso regressions are doing, familiarize yourself with the `glmnet` package's documentation. It is a thoroughly tested and well-established package that provides efficient code to compute the penalized regression coefficients for Ridge and Lasso and for combinations, commonly called *Elastic Nets*.

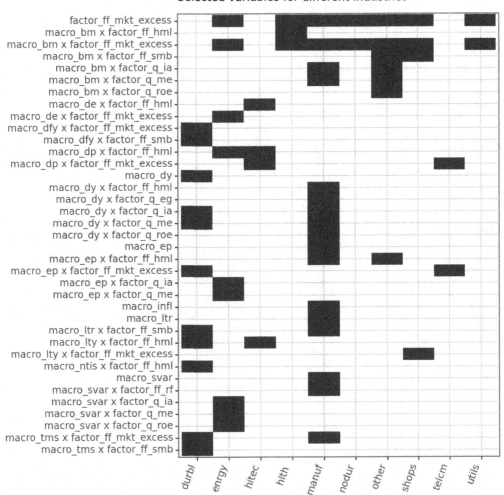

Figure 15.5: The figure shows selected variables for different industries. Dark areas indicate that the estimated Lasso regression coefficient is not set to zero. White fields show which variables get assigned a value of exactly zero.

# 16

# *Option Pricing via Machine Learning*

This chapter covers machine learning methods in option pricing. First, we briefly introduce regression trees, random forests, and neural networks; these methods are advocated as highly flexible *universal approximators*, capable of recovering highly non-linear structures in the data. As the focus is on implementation, we leave a thorough treatment of the statistical underpinnings to other textbooks from authors with a real comparative advantage on these issues. We show how to implement random forests and deep neural networks with tidy principles using `scikit-learn`.

Machine learning (ML) is seen as a part of artificial intelligence. ML algorithms build a model based on training data in order to make predictions or decisions without being explicitly programmed to do so. While ML can be specified along a vast array of different branches, this chapter focuses on so-called supervised learning for regressions. The basic idea of supervised learning algorithms is to build a mathematical model for data that contains both the inputs and the desired outputs. In this chapter, we apply well-known methods such as random forests and neural networks to a simple application in option pricing. More specifically, we create an artificial dataset of option prices for different values based on the Black-Scholes pricing equation for call options. Then, we train different models to *learn* how to price call options without prior knowledge of the theoretical underpinnings of the famous option pricing equation by Black and Scholes (1973).

Throughout this chapter, we need the following Python packages.

```python
import pandas as pd
import numpy as np

from plotnine import *
from itertools import product
from scipy.stats import norm
from sklearn.model_selection import train_test_split
from sklearn.pipeline import Pipeline
from sklearn.compose import ColumnTransformer
from sklearn.preprocessing import StandardScaler
from sklearn.ensemble import RandomForestRegressor
from sklearn.neural_network import MLPRegressor
from sklearn.preprocessing import PolynomialFeatures
from sklearn.linear_model import Lasso
```

## 16.1 Regression Trees and Random Forests

Regression trees are a popular ML approach for incorporating multiway predictor interactions. In Finance, regression trees are gaining popularity, also in the context of asset pricing (see, e.g., Bryzgalova et al., 2022). Trees possess a logic that departs markedly from traditional regressions. Trees are designed to find groups of observations that behave similarly to each other. A tree *grows* in a sequence of steps. At each step, a new *branch* sorts the data leftover from the preceding step into bins based on one of the predictor variables. This sequential branching slices the space of predictors into partitions and approximates the unknown function $f(x)$ which yields the relation between the predictors $x$ and the outcome variable $y$ with the average value of the outcome variable within each partition. For a more thorough treatment of regression trees, we refer to Coqueret and Guida (2020).

Formally, we partition the predictor space into $J$ non-overlapping regions, $R_1, R_2, \dots, R_J$. For any predictor $x$ that falls within region $R_j$, we estimate $f(x)$ with the average of the training observations, $\hat{y}_i$, for which the associated predictor $x_i$ is also in $R_j$. Once we select a partition $x$ to split in order to create the new partitions, we find a predictor $j$ and value $s$ that define two new partitions, called $R_1(j, s)$ and $R_2(j, s)$, which split our observations in the current partition by asking if $x_j$ is bigger than $s$:

$$R_1(j, s) = \{x \mid x_j < s\} \text{ and } R_2(j, s) = \{x \mid x_j \geq s\}. \tag{16.1}$$

To pick $j$ and $s$, we find the pair that minimizes the residual sum of square (RSS):

$$\sum_{i:\, x_i \in R_1(j,s)} (y_i - \hat{y}_{R_1})^2 + \sum_{i:\, x_i \in R_2(j,s)} (y_i - \hat{y}_{R_2})^2 \tag{16.2}$$

As in Chapter 15 in the context of penalized regressions, the first relevant question is: What are the hyperparameter decisions? Instead of a regularization parameter, trees are fully determined by the number of branches used to generate a partition (sometimes one specifies the minimum number of observations in each final branch instead of the maximum number of branches).

Models with a single tree may suffer from high predictive variance. Random forests address these shortcomings of decision trees. The goal is to improve the predictive performance and reduce instability by averaging multiple regression trees. A forest basically implies creating many regression trees and averaging their predictions. To assure that the individual trees are not the same, we use a bootstrap to induce randomness. More specifically, we build $B$ decision trees $T_1, \dots, T_B$ using the training sample. For that purpose, we randomly select features to be included in the building of each tree. For each observation in the test set, we then form a prediction $\hat{y} = \frac{1}{B} \sum_{i=1}^{B} \hat{y}_{T_i}$.

## 16.2 Neural Networks

Roughly speaking, neural networks propagate information from an input layer, through one or multiple hidden layers, to an output layer. While the number of units (neurons) in the input layer is equal to the dimension of the predictors, the output layer usually consists of one neuron (for regression) or multiple neurons for classification. The output layer predicts the future data, similar to the fitted value in a regression analysis. Neural networks have theoretical underpinnings as *universal approximators* for any smooth predictive association (Hornik, 1991). Their complexity, however, ranks neural networks among the least transparent, least interpretable, and most highly parameterized ML tools. In finance, applications of neural networks can be found in many different contexts, e.g., Avramov et al. (2022), Chen et al. (2023), and Gu et al. (2020).

Each neuron applies a non-linear *activation function f* to its aggregated signal before sending its output to the next layer

$$x_k^l = f\left(\theta_0^k + \sum_{j=1}^{N^l} z_j \theta_{l,j}^k\right) \tag{16.3}$$

Here, $\theta$ are the parameters to fit, $N^l$ denotes the number of units (a hyperparameter to tune), and $z_j$ are the input variables which can be either the raw data or, in the case of multiple chained layers, the outcome from a previous layer $z_j = x_k - 1$. While the easiest case with $f(x) = \alpha + \beta x$ resembles linear regression, typical activation functions are sigmoid (i.e., $f(x) = (1 + e^{-x})^{-1}$) or ReLu (i.e., $f(x) = max(x, 0)$).

Neural networks gain their flexibility from chaining multiple layers together. Naturally, this imposes many degrees of freedom on the network architecture for which no clear theoretical guidance exists. The specification of a neural network requires, at a minimum, a stance on depth (number of hidden layers), the activation function, the number of neurons, the connection structure of the units (dense or sparse), and the application of regularization techniques to avoid overfitting. Finally, *learning* means to choose optimal parameters relying on numerical optimization, which often requires specifying an appropriate learning rate. Despite these computational challenges, implementation in Python is not tedious at all because we can use the API to `TensorFlow`.

---

## 16.3   Option Pricing

To apply ML methods in a relevant field of finance, we focus on option pricing. The application in its core is taken from Hull (2020). In its most basic form, call options give the owner the right but not the obligation to buy a specific stock (the underlying) at a specific price (the strike price $K$) at a specific date (the exercise date $T$). The Black-Scholes price (Black and Scholes, 1973) of a call option for a non-dividend-paying underlying stock is given by

$$C(S, T) = \Phi(d_1)S - \Phi(d_2)Ke^{-rT}$$
$$d_1 = \frac{1}{\sigma\sqrt{T}}\left(\ln\left(\frac{S}{K}\right) + \left(r_f + \frac{\sigma^2}{2}\right)T\right) \tag{16.4}$$
$$d_2 = d_1 - \sigma\sqrt{T}$$

where $C(S, T)$ is the price of the option as a function of today's stock price of the underlying, $S$, with time to maturity $T$, $r_f$ is the risk-free interest rate, and $\sigma$ is the volatility of the underlying stock return. $\Phi$ is the cumulative distribution function of a standard normal random variable.

The Black-Scholes equation provides a way to compute the arbitrage-free price of a call option once the parameters $S, K, r_f, T$, and $\sigma$ are specified (arguably, in a realistic context, all parameters are easy to specify except for $\sigma$ which has to be estimated). A simple R function allows computing the price as we do below.

```python
def black_scholes_price(S, K, r, T, sigma):
    """Calculate Black Scholes option price."""

    d1 = (np.log(S/K)+(r+sigma**2/2)*T)/(sigma*np.sqrt(T))
    d2 = d1-sigma*np.sqrt(T)
    price = S*norm.cdf(d1)-K*np.exp(-r*T)*norm.cdf(d2)

    return price
```

## 16.4   Learning Black-Scholes

We illustrate the concept of ML by showing how ML methods *learn* the Black-Scholes equation after observing some different specifications and corresponding prices without us revealing the exact pricing equation.

### 16.4.1   Data simulation

To that end, we start with simulated data. We compute option prices for call options for a grid of different combinations of times to maturity (`T`), risk-free rates (`r`), volatilities (`sigma`), strike prices (`K`), and current stock prices (`S`). In the code below, we add an idiosyncratic error term to each observation such that the prices considered do not exactly reflect the values implied by the Black-Scholes equation.

In order to keep the analysis reproducible, we use `np.random.seed()`. A random seed specifies the start point when a computer generates a random number sequence and ensures that our simulated data is the same across different machines.

```python
random_state = 42
np.random.seed(random_state)


S = np.arange(40, 61)
K = np.arange(20, 91)
r = np.arange(0, 0.051, 0.01)
T = np.arange(3/12, 2.01, 1/12)
sigma = np.arange(0.1, 0.81, 0.1)

option_prices = pd.DataFrame(
  product(S, K, r, T, sigma),
  columns=["S", "K", "r", "T", "sigma"]
)


option_prices["black_scholes"] = black_scholes_price(
  option_prices["S"].values,
  option_prices["K"].values,
  option_prices["r"].values,
  option_prices["T"].values,
```

```
    option_prices["sigma"].values
)

option_prices = (option_prices
  .assign(
    observed_price=lambda x: (
      x["black_scholes"] + np.random.normal(scale=0.15)
    )
  )
)
```

The code above generates more than 1.5 million random parameter constellations (in the definition of the `option_prices` dataframe). For each of these values, the *true* prices reflecting the Black-Scholes model are given and a random innovation term *pollutes* the observed prices. The intuition of this application is simple: the simulated data provides many observations of option prices, by using the Black-Scholes equation we can evaluate the actual predictive performance of a ML method, which would be hard in a realistic context where the actual arbitrage-free price would be unknown.

Next, we split the data into a training set (which contains 1 percent of all the observed option prices) and a test set that will only be used for the final evaluation. Note that the entire grid of possible combinations contains `python len(option_prices.columns)` different specifications. Thus, the sample to learn the Black-Scholes price contains only 31,489 observations and is therefore relatively small.

```
train_data, test_data = train_test_split(
  option_prices,
  test_size=0.01, random_state=random_state
)
```

We process the training dataset further before we fit the different ML models. We define a `ColumnTransformer()` that defines all processing steps for that purpose. For our specific case, we want to explain the observed price by the five variables that enter the Black-Scholes equation. The *true* price (stored in column `black_scholes`) should obviously not be used to fit the model. The recipe also reflects that we standardize all predictors via `StandardScaler()` to ensure that each variable exhibits a sample average of zero and a sample standard deviation of one.

```
preprocessor = ColumnTransformer(
  transformers=[(
    "normalize_predictors",
    StandardScaler(),
    ["S", "K", "r", "T", "sigma"]
  )],
  remainder="drop"
)
```

### 16.4.2 Single layer networks and random forests

Next, we show how to fit a neural network to the data. The function `MLPRegressor()` from the package `scikit-learn` provides the functionality to initialize a single layer, feed-forward neural network. The specification below defines a single layer feed-forward neural network with ten hidden units. We set the number of training iterations to `max_iter=1000`.

```
max_iter = 1000

nnet_model = MLPRegressor(
  hidden_layer_sizes=10,
  max_iter=max_iter,
  random_state=random_state
)
```

We can follow the straightforward workflow as in the chapter before: define a workflow, equip it with the recipe, and specify the associated model. Finally, fit the model with the training data.

```
nnet_pipeline = Pipeline([
  ("preprocessor", preprocessor),
  ("regressor", nnet_model)
])

nnet_fit = nnet_pipeline.fit(
  train_data.drop(columns=["observed_price"]),
  train_data.get("observed_price")
)
```

One word of caution regarding the training of Neural networks: For illustrative purposes we sequentially update the parameters by reiterating through the data 1,000 times (`max_iter=1000`). Typically, however, early stopping rules are advised that aim to interrupt the process of updating parameters as soon as the predictive performance on the validation test seems to deteriorate. A detailed discussion of these details in the implementation would go beyond the scope of this book.

Once you are familiar with the `scikit-learn` workflow, it is a piece of cake to fit other models. For instance, the model below initializes a random forest with 50 trees contained in the ensemble, where we require at least 2000 observations in a node. The random forests are trained using the function `RandomForestRegressor()`.

```
rf_model = RandomForestRegressor(
  n_estimators=50,
  min_samples_leaf=2000,
  random_state=random_state
)
```

Fitting the model follows exactly the same convention as for the neural network before.

```
rf_pipeline = Pipeline([
  ("preprocessor", preprocessor),
  ("regressor", rf_model)
])

rf_fit = rf_pipeline.fit(
  train_data.drop(columns=["observed_price"]),
  train_data.get("observed_price")
)
```

### 16.4.3   Deep neural networks

A deep neural network is a neural network with multiple layers between the input and output layers. By chaining multiple layers together, more complex structures can be represented with fewer parameters than simple shallow (one-layer) networks as the one implemented above. For instance, image or text recognition are typical tasks where deep neural networks are used (for applications of deep neural networks in finance, see, for instance, Jiang et al., 2023; Jensen et al., 2022). The following code chunk implements a deep neural network with three hidden layers of size ten each and logistic activation functions.

```
deepnnet_model = MLPRegressor(
  hidden_layer_sizes=(10, 10, 10),
  activation="logistic",
  solver="lbfgs",
  max_iter=max_iter,
  random_state=random_state
)

deepnnet_pipeline = Pipeline([
  ("preprocessor", preprocessor),
  ("regressor", deepnnet_model)
])

deepnnet_fit = deepnnet_pipeline.fit(
  train_data.drop(columns=["observed_price"]),
  train_data.get("observed_price")
)
```

### 16.4.4   Universal approximation

Before we evaluate the results, we implement one more model. In principle, any non-linear function can also be approximated by a linear model containing the input variables' polynomial expansions. To illustrate this, we include polynomials up to the fifth degree of each predictor and then add all possible pairwise interaction terms. We fit a Lasso regression model with a pre-specified penalty term (consult Chapter 15 on how to tune the model hyperparameters).

```
lm_pipeline = Pipeline([
  ("polynomial", PolynomialFeatures(degree=5,
                                    interaction_only=False,
                                    include_bias=True)),
  ("scaler", StandardScaler()),
  ("regressor", Lasso(alpha=0.01))
])

lm_fit = lm_pipeline.fit(
  train_data.get(["S", "K", "r", "T", "sigma"]),
  train_data.get("observed_price")
)
```

## 16.5 Prediction Evaluation

Finally, we collect all predictions to compare the *out-of-sample* prediction error evaluated on 10,000 new data points.

```
test_X = test_data.get(["S", "K", "r", "T", "sigma"])
test_y = test_data.get("observed_price")

predictive_performance = (pd.concat(
    [test_data.reset_index(drop=True),
     pd.DataFrame({"Random forest": rf_fit.predict(test_X),
                   "Single layer": nnet_fit.predict(test_X),
                   "Deep NN": deepnnet_fit.predict(test_X),
                   "Lasso": lm_fit.predict(test_X)})
    ], axis=1)
  .melt(
    id_vars=["S", "K", "r", "T", "sigma",
             "black_scholes", "observed_price"],
    var_name="Model",
    value_name="Predicted"
  )
  .assign(
    moneyness=lambda x: x["S"]-x["K"],
    pricing_error=lambda x: np.abs(x["Predicted"]-x["black_scholes"])
  )
)
```

In the lines above, we use each of the fitted models to generate predictions for the entire test dataset of option prices. We evaluate the absolute pricing error as one possible measure of pricing accuracy, defined as the absolute value of the difference between predicted option price and the theoretical correct option price from the Black-Scholes model. We show the results graphically in Figure 16.1.

```
predictive_performance_plot = (
  ggplot(predictive_performance,
         aes(x="moneyness", y="pricing_error")) +
  geom_point(alpha=0.05) +
  facet_wrap("Model") +
  labs(x="Moneyness (S - K)", y="Absolut prediction error (USD)",
       title="Prediction errors of call options for different models") +
  theme(legend_position="")
)
predictive_performance_plot.draw()
```

The results can be summarized as follows:

1. All ML methods seem to be able to price call options after observing the training test set.
2. Random forest and the Lasso seem to perform consistently worse in predicting option prices than the neural networks.

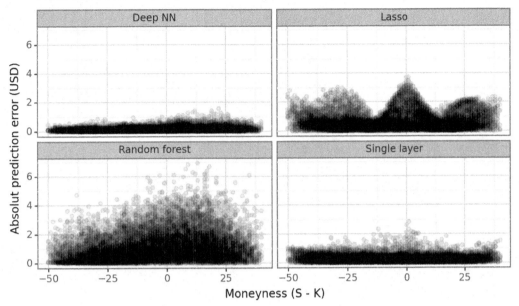

Figure 16.1: The figure shows absolut prediction error in USD for the different fitted methods. The prediction error is evaluated on a sample of call options that were not used for training.

3. For random forest and Lasso, the average prediction errors increase for far in-the-money options.
4. The increased complexity of the deep neural network relative to the single-layer neural network results in lower prediction errors.

## 16.6  Exercises

1. Write a function that takes y and a matrix of predictors X as inputs and returns a characterization of the relevant parameters of a regression tree with one branch.
2. Create a function that creates predictions for a new matrix of predictors newX based on the estimated regression tree.

# Part V

# Portfolio Optimization

# 17

## *Parametric Portfolio Policies*

In this chapter, we apply different portfolio performance measures to evaluate and compare portfolio allocation strategies. For this purpose, we introduce a direct way to estimate optimal portfolio weights for large-scale cross-sectional applications. More precisely, the approach of Brandt et al. (2009) proposes to parametrize the optimal portfolio weights as a function of stock characteristics instead of estimating the stock's expected return, variance, and covariances with other stocks in a prior step. We choose weights as a function of characteristics that maximize the expected utility of the investor. This approach is feasible for large portfolio dimensions (such as the entire CRSP universe) and has been proposed by Brandt et al. (2009). See the review paper by Brandt (2010) for an excellent treatment of related portfolio choice methods.

The current chapter relies on the following set of Python packages:

```python
import pandas as pd
import numpy as np
import sqlite3
import statsmodels.formula.api as smf

from itertools import product, starmap
from scipy.optimize import minimize
```

Compared to previous chapters, we introduce the `scipy.optimize` module from the `scipy` (Virtanen et al., 2020) for solving optimization problems.

## 17.1 Data Preparation

To get started, we load the monthly CRSP file, which forms our investment universe. We load the data from our SQLite database introduced in Chapter 3 and Chapter 4.

```python
tidy_finance = sqlite3.connect(database="data/tidy_finance_python.sqlite")

crsp_monthly = (pd.read_sql_query(
    sql=("SELECT permno, month, ret_excess, mktcap, mktcap_lag "
        "FROM crsp_monthly"),
    con=tidy_finance,
    parse_dates={"month"})
  .dropna()
)
```

To evaluate the performance of portfolios, we further use monthly market returns as a benchmark to compute CAPM alphas.

```
factors_ff_monthly = pd.read_sql_query(
  sql="SELECT month, mkt_excess FROM factors_ff3_monthly",
  con=tidy_finance,
  parse_dates={"month"}
)
```

Next, we retrieve some stock characteristics that have been shown to have an effect on the expected returns or expected variances (or even higher moments) of the return distribution. In particular, we record the lagged one-year return momentum (`momentum_lag`), defined as the compounded return between months $t - 13$ and $t - 2$ for each firm. In finance, momentum is the empirically observed tendency for rising asset prices to rise further and falling prices to keep falling (Jegadeesh and Titman, 1993). The second characteristic is the firm's market equity (`size_lag`), defined as the log of the price per share times the number of shares outstanding (Banz, 1981). To construct the correct lagged values, we use the approach introduced in Chapter 4.

```
crsp_monthly_lags = (crsp_monthly
  .assign(month=lambda x: x["month"]+pd.DateOffset(months=13))
  .get(["permno", "month", "mktcap"])
)

crsp_monthly = (crsp_monthly
  .merge(crsp_monthly_lags,
         how="inner", on=["permno", "month"], suffixes=["", "_13"])
)

data_portfolios = (crsp_monthly
  .assign(
    momentum_lag=lambda x: x["mktcap_lag"]/x["mktcap_13"],
    size_lag=lambda x: np.log(x["mktcap_lag"])
  )
  .dropna(subset=["momentum_lag", "size_lag"])
)
```

## 17.2   Parametric Portfolio Policies

The basic idea of parametric portfolio weights is as follows. Suppose that at each date $t$, we have $N_t$ stocks in the investment universe, where each stock $i$ has a return of $r_{i,t+1}$ and is associated with a vector of firm characteristics $x_{i,t}$ such as time-series momentum or the market capitalization. The investor's problem is to choose portfolio weights $w_{i,t}$ to maximize the expected utility of the portfolio return:

$$\max_{\omega} E_t\left(u(r_{p,t+1})\right) = E_t\left[u\left(\sum_{i=1}^{N_t} \omega_{i,t} \cdot r_{i,t+1}\right)\right] \tag{17.1}$$

where $u(\cdot)$ denotes the utility function.

Where do the stock characteristics show up? We parameterize the optimal portfolio weights as a function of the stock characteristic $x_{i,t}$ with the following linear specification for the portfolio weights:

$$\omega_{i,t} = \bar{\omega}_{i,t} + \frac{1}{N_t}\theta'\hat{x}_{i,t}, \tag{17.2}$$

where $\bar{\omega}_{i,t}$ is a stock's weight in a benchmark portfolio (we use the value-weighted or naive portfolio in the application below), $\theta$ is a vector of coefficients which we are going to estimate, and $\hat{x}_{i,t}$ are the characteristics of stock $i$, cross-sectionally standardized to have zero mean and unit standard deviation.

Intuitively, the portfolio strategy is a form of active portfolio management relative to a performance benchmark. Deviations from the benchmark portfolio are derived from the individual stock characteristics. Note that by construction, the weights sum up to one as $\sum_{i=1}^{N_t}\hat{x}_{i,t} = 0$ due to the standardization. Moreover, the coefficients are constant across assets and over time. The implicit assumption is that the characteristics fully capture all aspects of the joint distribution of returns that are relevant for forming optimal portfolios.

We first implement cross-sectional standardization for the entire CRSP universe. We also keep track of (lagged) relative market capitalization `relative_mktcap`, which will represent the value-weighted benchmark portfolio, while **n** denotes the number of traded assets $N_t$, which we use to construct the naive portfolio benchmark.

```
data_portfolios = (data_portfolios
  .groupby("month")
  .apply(lambda x: x.assign(
     relative_mktcap=x["mktcap_lag"]/x["mktcap_lag"].sum()
   )
 )
  .reset_index(drop=True)
  .set_index("month")
  .groupby(level="month")
  .transform(
    lambda x: (x-x.mean())/x.std() if x.name.endswith("lag") else x
  )
  .reset_index()
  .drop(["mktcap_lag"], axis=1)
)
```

## 17.3 Computing Portfolio Weights

Next, we move on to identify optimal choices of $\theta$. We rewrite the optimization problem together with the weight parametrization and can then estimate $\theta$ to maximize the objective function based on our sample

$$E_t\left(u(r_{p,t+1})\right) = \frac{1}{T}\sum_{t=0}^{T-1}u\left(\sum_{i=1}^{N_t}\left(\bar{\omega}_{i,t} + \frac{1}{N_t}\theta'\hat{x}_{i,t}\right)r_{i,t+1}\right). \tag{17.3}$$

The allocation strategy is straightforward because the number of parameters to estimate is small. Instead of a tedious specification of the $N_t$ dimensional vector of expected returns and the $N_t(N_t + 1)/2$ free elements of the covariance matrix, all we need to focus on in our application is the vector $\theta$. $\theta$ contains only two elements in our application: the relative deviation from the benchmark due to *size* and *momentum*.

To get a feeling for the performance of such an allocation strategy, we start with an arbitrary initial vector $\theta_0$. The next step is to choose $\theta$ optimally to maximize the objective function. We automatically detect the number of parameters by counting the number of columns with lagged values. Note that the value for $\theta$ of 1.5 is an arbitrary choice.

```
lag_columns = [i for i in data_portfolios.columns if "lag" in i]
n_parameters = len(lag_columns)
theta = pd.DataFrame({"theta": [1.5]*n_parameters}, index=lag_columns)
```

The function `compute_portfolio_weights()` below computes the portfolio weights $\bar{\omega}_{i,t} + \frac{1}{N_t}\theta'\hat{x}_{i,t}$ according to our parametrization for a given value $\theta_0$. Everything happens within a single pipeline. Hence, we provide a short walk-through.

We first compute `characteristic_tilt`, the tilting values $\frac{1}{N_t}\theta'\hat{x}_{i,t}$ which resemble the deviation from the benchmark portfolio. Next, we compute the benchmark portfolio `weight_benchmark`, which can be any reasonable set of portfolio weights. In our case, we choose either the value or equal-weighted allocation. `weight_tilt` completes the picture and contains the final portfolio weights `weight_tilt = weight_benchmark + characteristic_tilt`, which deviate from the benchmark portfolio depending on the stock characteristics.

The final few lines go a bit further and implement a simple version of a no-short sale constraint. While it is generally not straightforward to ensure portfolio weight constraints via parameterization, we simply normalize the portfolio weights such that they are enforced to be positive. Finally, we make sure that the normalized weights sum up to one again:

$$\omega_{i,t}^+ = \frac{\max(0, \omega_{i,t})}{\sum_{j=1}^{N_t} \max(0, \omega_{i,t})}. \tag{17.4}$$

The following function computes the optimal portfolio weights in the way just described.

```
def compute_portfolio_weights(theta,
                              data,
                              value_weighting=True,
                              allow_short_selling=True):
    """Compute portfolio weights for different strategies."""

    lag_columns = [i for i in data.columns if "lag" in i]
    theta = pd.DataFrame(theta, index=lag_columns)

    data = (data
      .groupby("month")
      .apply(lambda x: x.assign(
          characteristic_tilt=x[theta.index] @ theta / x.shape[0]
        )
      )
      .reset_index(drop=True)
      .assign(
        weight_benchmark=lambda x:
```

```
      x["relative_mktcap"] if value_weighting else 1/x.shape[0],
    weight_tilt=lambda x:
      x["weight_benchmark"] + x["characteristic_tilt"]
  )
  .drop(columns=["characteristic_tilt"])
)

if not allow_short_selling:
    data = (data
      .assign(weight_tilt=lambda x: np.maximum(0, x["weight_tilt"]))
    )

data = (data
  .groupby("month")
  .apply(lambda x: x.assign(
    weight_tilt=lambda x: x["weight_tilt"]/x["weight_tilt"].sum()))
  .reset_index(drop=True)
)

return data
```

In the next step, we compute the portfolio weights for the arbitrary vector $\theta_0$. In the example below, we use the value-weighted portfolio as a benchmark and allow negative portfolio weights.

```
weights_crsp = compute_portfolio_weights(
  theta,
  data_portfolios,
  value_weighting=True,
  allow_short_selling=True
)
```

## 17.4   Portfolio Performance

Are the computed weights optimal in any way? Most likely not, as we picked $\theta_0$ arbitrarily. To evaluate the performance of an allocation strategy, one can think of many different approaches. In their original paper, Brandt et al. (2009) focus on a simple evaluation of the hypothetical utility of an agent equipped with a power utility function

$$u_\gamma(r) = \frac{(1 + r)^{(1-\gamma)}}{1 - \gamma}, \tag{17.5}$$

where $\gamma$ is the risk aversion factor.

```
def power_utility(r, gamma=5):
    """Calculate power utility for given risk aversion."""

    utility = ((1+r)**(1-gamma))/(1-gamma)
```

**`return` utility**

We want to note that Gehrig et al. (2020) warn that, in the leading case of constant relative risk aversion (CRRA), strong assumptions on the properties of the returns, the variables used to implement the parametric portfolio policy, and the parameter space are necessary to obtain a well-defined optimization problem.

No doubt, there are many other ways to evaluate a portfolio. The function below provides a summary of all kinds of interesting measures that can be considered relevant. Do we need all these evaluation measures? It depends: The original paper by Brandt et al. (2009) only cares about the expected utility to choose $\theta$. However, if you want to choose optimal values that achieve the highest performance while putting some constraints on your portfolio weights, it is helpful to have everything in one function.

```python
def evaluate_portfolio(weights_data,
                       full_evaluation=True,
                       capm_evaluation=True,
                       length_year=12):
    """Calculate portfolio evaluation measures."""
    evaluation = (weights_data
        .groupby("month")
        .apply(lambda x: pd.Series(
          np.average(x[["ret_excess", "ret_excess"]],
                     weights=x[["weight_tilt", "weight_benchmark"]],
                     axis=0),
          ["return_tilt", "return_benchmark"])
        )
        .reset_index()
        .melt(id_vars="month", var_name="model",
              value_vars=["return_tilt", "return_benchmark"],
              value_name="portfolio_return")
        .assign(model=lambda x: x["model"].str.replace("return_", ""))
    )

    evaluation_stats = (evaluation
        .groupby("model")["portfolio_return"]
        .aggregate([
          ("Expected utility", lambda x: np.mean(power_utility(x))),
          ("Average return", lambda x: np.mean(length_year*x)*100),
          ("SD return", lambda x: np.std(x)*np.sqrt(length_year)*100),
          ("Sharpe ratio", lambda x: (np.mean(x)/np.std(x)*
                                      np.sqrt(length_year)))
        ])
    )

    if capm_evaluation:
        evaluation_capm = (evaluation
            .merge(factors_ff_monthly, how="left", on="month")
            .groupby("model")
            .apply(lambda x:
              smf.ols(formula="portfolio_return ~ 1 + mkt_excess", data=x)
```

```
                    .fit().params
            )
            .rename(columns={"const": "CAPM alpha",
                              "mkt_excess": "Market beta"})
        )
        evaluation_stats = (evaluation_stats
          .merge(evaluation_capm, how="left", on="model")
        )

    if full_evaluation:
        evaluation_weights = (weights_data
          .melt(id_vars="month", var_name="model",
                value_vars=["weight_benchmark", "weight_tilt"],
                value_name="weight")
          .groupby(["model", "month"])["weight"]
          .aggregate([
            ("Mean abs. weight", lambda x: np.mean(abs(x))),
            ("Max. weight", lambda x: max(x)),
            ("Min. weight", lambda x: min(x)),
            ("Avg. sum of neg. weights", lambda x: -np.sum(x[x < 0])),
            ("Avg. share of neg. weights", lambda x: np.mean(x < 0))
          ])
          .reset_index()
          .drop(columns=["month"])
          .groupby(["model"])
          .aggregate(lambda x: np.average(x)*100)
          .reset_index()
          .assign(model=lambda x: x["model"].str.replace("weight_", ""))
        )

        evaluation_stats = (evaluation_stats
          .merge(evaluation_weights, how="left", on="model")
          .set_index("model")
        )

    evaluation_stats = (evaluation_stats
      .transpose()
      .rename_axis(columns=None)
    )

    return evaluation_stats
```

Let us take a look at the different portfolio strategies and evaluation measures.

```
evaluate_portfolio(weights_crsp).round(2)
```

|                  | benchmark | tilt  |
| ---------------- | --------- | ----- |
| Expected utility | -0.25     | -0.26 |
| Average return   | 6.65      | 0.19  |
| SD return        | 15.46     | 21.11 |

|                          | benchmark | tilt  |
| ------------------------ | --------- | ----- |
| Sharpe ratio             | 0.43      | 0.01  |
| Intercept                | 0.00      | -0.01 |
| Market beta              | 0.99      | 0.95  |
| Mean abs. weight         | 0.03      | 0.08  |
| Max. weight              | 4.05      | 4.22  |
| Min. weight              | 0.00      | -0.17 |
| Avg. sum of neg. weights | 0.00      | 78.03 |
| Avg. share of neg. weights | 0.00    | 49.08 |

The value-weighted portfolio delivers an annualized return of more than six percent and clearly outperforms the tilted portfolio, irrespective of whether we evaluate expected utility, the Sharpe ratio, or the CAPM alpha. We can conclude the market beta is close to one for both strategies (naturally almost identically one for the value-weighted benchmark portfolio). When it comes to the distribution of the portfolio weights, we see that the benchmark portfolio weight takes less extreme positions (lower average absolute weights and lower maximum weight). By definition, the value-weighted benchmark does not take any negative positions, while the tilted portfolio also takes short positions.

## 17.5   Optimal Parameter Choice

Next, we move to a choice of $\theta$ that actually aims to improve some (or all) of the performance measures. We first define the helper function `compute_objective_function()`, which we then pass to an optimizer.

```python
def objective_function(theta,
                       data,
                       objective_measure="Expected utility",
                       value_weighting=True,
                       allow_short_selling=True):
    """Define portfolio objective function."""

    processed_data = compute_portfolio_weights(
      theta, data, value_weighting, allow_short_selling
    )

    objective_function = evaluate_portfolio(
      processed_data,
      capm_evaluation=False,
      full_evaluation=False
    )

    objective_function = -objective_function.loc[objective_measure, "tilt"]

    return objective_function
```

You may wonder why we return the negative value of the objective function. This is simply due to the common convention for optimization procedures to search for minima as a default. By minimizing the negative value of the objective function, we get the maximum value as a result. In its most basic form, Python optimization uses the function `minimize()`. As main inputs, the function requires an initial guess of the parameters and the objective function to minimize. Now, we are fully equipped to compute the optimal values of $\theta$, which maximize the hypothetical expected utility of the investor.

```python
optimal_theta = minimize(
  fun=objective_function,
  x0=[1.5]*n_parameters,
  args=(data_portfolios, "Expected utility", True, True),
  method="Nelder-Mead",
  tol=1e-2
)

(pd.DataFrame(
  optimal_theta.x,
  columns=["Optimal theta"],
  index=["momentum_lag", "size_lag"]).T.round(3)
)
```

|  | momentum_lag | size_lag |
|---|---|---|
| Optimal theta | 0.344 | -1.827 |

The resulting values of $\hat{\theta}$ are easy to interpret: intuitively, expected utility increases by tilting weights from the value-weighted portfolio toward smaller stocks (negative coefficient for size) and toward past winners (positive value for momentum). Both findings are in line with the well-documented size effect (Banz, 1981) and the momentum anomaly (Jegadeesh and Titman, 1993).

## 17.6 More Model Specifications

How does the portfolio perform for different model specifications? For this purpose, we compute the performance of a number of different modeling choices based on the entire CRSP sample. The next code chunk performs all the heavy lifting.

```python
def evaluate_optimal_performance(data,
                                 objective_measure="Expected utility",
                                 value_weighting=True,
                                 allow_short_selling=True):
    """Calculate optimal portfolio performance."""

    optimal_theta = minimize(
      fun=objective_function,
      x0=[1.5]*n_parameters,
      args=(data, objective_measure, value_weighting, allow_short_selling),
```

```
    method="Nelder-Mead",
    tol=10e-2
).x

processed_data = compute_portfolio_weights(
    optimal_theta, data,
    value_weighting, allow_short_selling
)

portfolio_evaluation = evaluate_portfolio(processed_data)

weight_text = "VW" if value_weighting else "EW"
short_text = "" if allow_short_selling else " (no s.)"

strategy_name_dict = {
    "benchmark": weight_text,
    "tilt": f"{weight_text} Optimal{short_text}"
}

portfolio_evaluation.columns = [
    strategy_name_dict[i] for i in portfolio_evaluation.columns
]

return(portfolio_evaluation)
```

Finally, we can compare the results. The table below shows summary statistics for all possible combinations: equal- or value-weighted benchmark portfolio, with or without short-selling constraints, and tilted toward maximizing expected utility.

```
data = [data_portfolios]
value_weighting = [True, False]
allow_short_selling = [True, False]
objective_measure = ["Expected utility"]

permutations = product(
    data, objective_measure,
    value_weighting, allow_short_selling
)
results = list(starmap(
    evaluate_optimal_performance,
    permutations
))
performance_table = (pd.concat(results, axis=1)
    .T.drop_duplicates().T.round(3)
)
performance_table.get(["EW", "VW"])
```

|                   | EW     | VW     |
|-------------------|--------|--------|
| Expected utility  | -0.251 | -0.250 |
| Average return    | 10.009 | 6.649  |

|                          | EW     | VW     |
| ------------------------ | ------ | ------ |
| SD return                | 20.352 | 15.462 |
| Sharpe ratio             | 0.492  | 0.430  |
| Intercept                | 0.002  | 0.000  |
| Market beta              | 1.125  | 0.994  |
| Mean abs. weight         | 0.000  | 0.030  |
| Max. weight              | 0.000  | 4.053  |
| Min. weight              | 0.000  | 0.000  |
| Avg. sum of neg. weights | 0.000  | 0.000  |
| Avg. share of neg. weights | 0.000 | 0.000 |

```
performance_table.get(["EW Optimal", "VW Optimal"])
```

|                          | EW Optimal | VW Optimal |
| ------------------------ | ---------- | ---------- |
| Expected utility         | -5.840     | -0.261     |
| Average return           | -4948.462  | 0.194      |
| SD return                | 14729.881  | 21.106     |
| Sharpe ratio             | -0.336     | 0.009      |
| Intercept                | -3.697     | -0.005     |
| Market beta              | -78.566    | 0.952      |
| Mean abs. weight         | 62.166     | 0.077      |
| Max. weight              | 1119.202   | 4.216      |
| Min. weight              | -221.305   | -0.174     |
| Avg. sum of neg. weights | 77916.229  | 78.025     |
| Avg. share of neg. weights | 51.878   | 49.076     |

```
performance_table.get(["EW Optimal (no s.)", "VW Optimal (no s.)"])
```

|                          | EW Optimal (no s.) | VW Optimal (no s.) |
| ------------------------ | ------------------ | ------------------ |
| Expected utility         | -0.252             | -0.250             |
| Average return           | 7.962              | 7.301              |
| SD return                | 19.126             | 16.700             |
| Sharpe ratio             | 0.416              | 0.437              |
| Intercept                | 0.000              | 0.000              |
| Market beta              | 1.136              | 1.055              |
| Mean abs. weight         | 0.030              | 0.030              |
| Max. weight              | 1.309              | 2.332              |
| Min. weight              | 0.000              | 0.000              |
| Avg. sum of neg. weights | 0.000              | 0.000              |
| Avg. share of neg. weights | 0.000            | 0.000              |

The results indicate that the average annualized Sharpe ratio of the equal-weighted portfolio exceeds the Sharpe ratio of the value-weighted benchmark portfolio. Nevertheless, starting with the weighted value portfolio as a benchmark and tilting optimally with respect to momentum and small stocks yields the highest Sharpe ratio across all specifications. Finally, imposing no short-sale constraints does not improve the performance of the portfolios in our application.

## 17.7   Exercises

1. How do the estimated parameters $\hat{\theta}$ and the portfolio performance change if your objective is to maximize the Sharpe ratio instead of the hypothetical expected utility?

2. The code above is very flexible in the sense that you can easily add new firm characteristics. Construct a new characteristic of your choice and evaluate the corresponding coefficient $\hat{\theta}_i$.

3. Tweak the function `optimal_theta()` such that you can impose additional performance constraints in order to determine $\hat{\theta}$, which maximizes expected utility under the constraint that the market beta is below 1.

4. Does the portfolio performance resemble a realistic out-of-sample backtesting procedure? Verify the robustness of the results by first estimating $\hat{\theta}$ based on *past data* only. Then, use more recent periods to evaluate the actual portfolio performance.

5. By formulating the portfolio problem as a statistical estimation problem, you can easily obtain standard errors for the coefficients of the weight function. Brandt et al. (2009) provide the relevant derivations in their paper in Equation (10). Implement a small function that computes standard errors for $\hat{\theta}$.

# 18

# Constrained Optimization and Backtesting

In this chapter, we conduct portfolio backtesting in a realistic setting by including transaction costs and investment constraints such as no-short-selling rules. We start with standard mean-variance efficient portfolios and introduce constraints in a step-by-step manner. To do so, we rely on numerical optimization procedures in Python. We conclude the chapter by providing an out-of-sample backtesting procedure for the different strategies that we introduce in this chapter.

Throughout this chapter, we use the following Python packages:

```python
import pandas as pd
import numpy as np
import sqlite3

from plotnine import *
from mizani.formatters import percent_format
from itertools import product
from scipy.stats import expon
from scipy.optimize import minimize
```

Compared to previous chapters, we introduce `expon` from `scipy.stats` to calculate exponential continuous random variables.

## 18.1 Data Preparation

We start by loading the required data from our SQLite database introduced in Chapter 3 and Chapter 4. For simplicity, we restrict our investment universe to the monthly Fama-French industry portfolio returns in the following application.

```python
tidy_finance = sqlite3.connect(database="data/tidy_finance_python.sqlite")

industry_returns = (pd.read_sql_query(
    sql="SELECT * FROM industries_ff_monthly",
    con=tidy_finance,
    parse_dates={"month"})
  .drop(columns=["month"])
)
```

## 18.2   Recap of Portfolio Choice

A common objective for portfolio optimization is to find mean-variance efficient portfolio weights, i.e., the allocation that delivers the lowest possible return variance for a given minimum level of expected returns. In the most extreme case, where the investor is only concerned about portfolio variance, they may choose to implement the minimum variance portfolio (MVP) weights which are given by the solution to

$$\omega_{\text{mvp}} = \arg\min \omega' \Sigma \omega \text{ s.t. } \omega' \iota = 1 \tag{18.1}$$

where $\Sigma$ is the $(N \times N)$ covariance matrix of the returns. The optimal weights $\omega_{\text{mvp}}$ can be found analytically and are $\omega_{\text{mvp}} = \frac{\Sigma^{-1}\iota}{\iota'\Sigma^{-1}\iota}$. In terms of code, the math is equivalent to the following chunk.

```
n_industries = industry_returns.shape[1]

mu = np.array(industry_returns.mean()).T
sigma = np.array(industry_returns.cov())
w_mvp = np.linalg.inv(sigma) @ np.ones(n_industries)
w_mvp = w_mvp/w_mvp.sum()

weights_mvp = pd.DataFrame({
  "Industry": industry_returns.columns.tolist(),
  "Minimum variance": w_mvp
})
weights_mvp.round(3)
```

|   | Industry | Minimum variance |
|---|----------|------------------|
| 0 | nodur    | 0.248            |
| 1 | durbl    | -0.012           |
| 2 | manuf    | 0.078            |
| 3 | enrgy    | 0.079            |
| 4 | hitec    | 0.008            |
| 5 | telcm    | 0.241            |
| 6 | shops    | 0.092            |
| 7 | hlth     | 0.160            |
| 8 | utils    | 0.470            |
| 9 | other    | -0.364           |

Next, consider an investor who aims to achieve minimum variance *given a required expected portfolio return* $\bar{\mu}$ such that she chooses

$$\omega_{\text{eff}}(\bar{\mu}) = \arg\min \omega' \Sigma \omega \text{ s.t. } \omega' \iota = 1 \text{ and } \omega' \mu \geq \bar{\mu}. \tag{18.2}$$

We leave it as an exercise below to show that the portfolio choice problem can equivalently be formulated for an investor with mean-variance preferences and risk aversion factor $\gamma$. That means the investor aims to choose portfolio weights as the solution to

$$\omega_{\gamma}^* = \arg\max \omega' \mu - \frac{\gamma}{2} \omega' \Sigma \omega \quad \text{s.t. } \omega' \iota = 1. \tag{18.3}$$

The solution to the optimal portfolio choice problem is:

$$\omega_\gamma^* = \frac{1}{\gamma}\left(\Sigma^{-1} - \frac{1}{\iota'\Sigma^{-1}\iota}\Sigma^{-1}\iota\iota'\Sigma^{-1}\right)\mu + \frac{1}{\iota'\Sigma^{-1}\iota}\Sigma^{-1}\iota. \tag{18.4}$$

To proof this statement, we refer to the derivations in Appendix B. Empirically, this classical solution imposes many problems. In particular, the estimates of $\mu$ are noisy over short horizons, the $(N \times N)$ matrix $\Sigma$ contains $N(N-1)/2$ distinct elements and thus, estimation error is huge. Seminal papers on the effect of ignoring estimation uncertainty, among others, are Brown (1976), Jobson and Korkie (1980), Jorion (1986), and Chopra and Ziemba (1993).

Even worse, if the asset universe contains more assets than available time periods $(N > T)$, the sample covariance matrix is no longer positive definite such that the inverse $\Sigma^{-1}$ does not exist anymore. To address estimation issues for vast-dimensional covariance matrices, regularization techniques (see, e.g., Ledoit and Wolf, 2003, 2004, 2012; Fan et al., 2008) and the parametric approach from the previous chapter are popular tools.

While the uncertainty associated with estimated parameters is challenging, the data-generating process is also unknown to the investor. In other words, model uncertainty reflects that it is ex-ante not even clear which parameters require estimation (for instance, if returns are driven by a factor model, selecting the universe of relevant factors imposes model uncertainty). Wang (2005) and Garlappi et al. (2007) provide theoretical analysis on optimal portfolio choice under model *and* estimation uncertainty. In the most extreme case, Pflug et al. (2012) shows that the naive portfolio, which allocates equal wealth to all assets, is the optimal choice for an investor averse to model uncertainty.

On top of the estimation uncertainty, *transaction costs* are a major concern. Rebalancing portfolios is costly, and, therefore, the optimal choice should depend on the investor's current holdings. In the presence of transaction costs, the benefits of reallocating wealth may be smaller than the costs associated with turnover. This aspect has been investigated theoretically, among others, for one risky asset by Magill and Constantinides (1976) and Davis and Norman (1990). Subsequent extensions to the case with multiple assets have been proposed by Balduzzi and Lynch (1999) and Balduzzi and Lynch (2000). More recent papers on empirical approaches that explicitly account for transaction costs include Gârleanu and Pedersen (2013), DeMiguel et al. (2014), and DeMiguel et al. (2015).

## 18.3  Estimation Uncertainty and Transaction Costs

The empirical evidence regarding the performance of a mean-variance optimization procedure in which you simply plug in some sample estimates $\hat{\mu}$ and $\hat{\Sigma}$ can be summarized rather briefly: mean-variance optimization performs poorly! The literature discusses many proposals to overcome these empirical issues. For instance, one may impose some form of regularization of $\Sigma$, rely on Bayesian priors inspired by theoretical asset pricing models (Kan and Zhou, 2007), or use high-frequency data to improve forecasting (Hautsch et al., 2015). One unifying framework that works easily, effectively (even for large dimensions), and is purely inspired by economic arguments is an ex-ante adjustment for transaction costs (Hautsch and Voigt, 2019).

Assume that returns are from a multivariate normal distribution with mean $\mu$ and variance-covariance matrix $\Sigma$, $N(\mu, \Sigma)$. Additionally, we assume quadratic transaction costs which penalize rebalancing such that

$$\nu\left(\omega_{t+1}, \omega_{t^+}, \beta\right) = \frac{\beta}{2}\left(\omega_{t+1} - \omega_{t^+}\right)'\left(\omega_{t+1} - \omega_{t^+}\right), \tag{18.5}$$

with cost parameter $\beta > 0$ and $\omega_{t^+} = \omega_t \circ (1 + r_t)/\iota'(\omega_t \circ (1 + r_t))$, where $\circ$ is the element-wise Hadamard product. $\omega_{t^+}$ denotes the portfolio weights just before rebalancing. Note that $\omega_{t^+}$ differs mechanically from $\omega_t$ due to the returns in the past period. Intuitively, transaction costs penalize portfolio performance when the portfolio is shifted from the current holdings $\omega_{t^+}$ to a new allocation $\omega_{t+1}$. In this setup, transaction costs do not increase linearly. Instead, larger rebalancing is penalized more heavily than small adjustments. Then, the optimal portfolio choice for an investor with mean variance preferences is

$$\begin{aligned} \omega_{t+1}^* &= \arg\max \omega'\mu - \nu_t(\omega, \omega_{t^+}, \beta) - \frac{\gamma}{2}\omega'\Sigma\omega \text{ s.t. } \iota'\omega = 1 \\ &= \arg\max \omega'\mu^* - \frac{\gamma}{2}\omega'\Sigma^*\omega \text{ s.t.} \iota'\omega = 1, \end{aligned} \tag{18.6}$$

where

$$\mu^* = \mu + \beta\omega_{t^+} \quad \text{and} \quad \Sigma^* = \Sigma + \frac{\beta}{\gamma}I_N. \tag{18.7}$$

As a result, adjusting for transaction costs implies a standard mean-variance optimal portfolio choice with adjusted return parameters $\Sigma^*$ and $\mu^*$:

$$\omega_{t+1}^* = \frac{1}{\gamma}\left(\Sigma^{*-1} - \frac{1}{\iota'\Sigma^{*-1}\iota}\Sigma^{*-1}\iota\iota'\Sigma^{*-1}\right)\mu^* + \frac{1}{\iota'\Sigma^{*-1}\iota}\Sigma^{*-1}\iota. \tag{18.8}$$

An alternative formulation of the optimal portfolio can be derived as follows:

$$\omega_{t+1}^* = \arg\max \omega'\left(\mu + \beta\left(\omega_{t^+} - \frac{1}{N}\iota\right)\right) - \frac{\gamma}{2}\omega'\Sigma^*\omega \text{ s.t. } \iota'\omega = 1. \tag{18.9}$$

The optimal weights correspond to a mean-variance portfolio, where the vector of expected returns is such that assets that currently exhibit a higher weight are considered as delivering a higher expected return.

---

## 18.4   Optimal Portfolio Choice

The function below implements the efficient portfolio weights in its general form, allowing for transaction costs (conditional on the holdings *before* reallocation). For $\beta = 0$, the computation resembles the standard mean-variance efficient framework. `gamma` denotes the coefficient of risk aversion $\gamma$, `beta` is the transaction cost parameter $\beta$ and `w_prev` are the weights before rebalancing $\omega_{t^+}$.

```python
def compute_efficient_weight(sigma,
                             mu,
                             gamma=2,
                             beta=0,
                             w_prev=np.ones(sigma.shape[1])/sigma.shape[1]):
    """Compute efficient portfolio weights."""
```

```
    n = sigma.shape[1]
    iota = np.ones(n)
    sigma_processed = sigma+(beta/gamma)*np.eye(n)
    mu_processed = mu+beta*w_prev

    sigma_inverse = np.linalg.inv(sigma_processed)

    w_mvp = sigma_inverse @ iota
    w_mvp = w_mvp/np.sum(w_mvp)
    w_opt = w_mvp+(1/gamma)*\
        (sigma_inverse-np.outer(w_mvp, iota) @ sigma_inverse) @ mu_processed

    return w_opt

w_efficient = compute_efficient_weight(sigma, mu)

weights_efficient = pd.DataFrame({
  "Industry": industry_returns.columns.tolist(),
  "Efficient portfolio": w_efficient
})
weights_efficient.round(3)
```

|   | Industry | Efficient portfolio |
|---|----------|---------------------|
| 0 | nodur    | 1.630               |
| 1 | durbl    | 0.090               |
| 2 | manuf    | -1.356              |
| 3 | enrgy    | 0.687               |
| 4 | hitec    | 0.333               |
| 5 | telcm    | -0.412              |
| 6 | shops    | 0.504               |
| 7 | hlth     | 0.402               |
| 8 | utils    | -0.219              |
| 9 | other    | -0.659              |

The portfolio weights above indicate the efficient portfolio for an investor with risk aversion coefficient $\gamma = 2$ in the absence of transaction costs. Some of the positions are negative, which implies short-selling, and most of the positions are rather extreme. For instance, a position of $-1$ implies that the investor takes a short position worth their entire wealth to lever long positions in other assets. What is the effect of transaction costs or different levels of risk aversion on the optimal portfolio choice? The following few lines of code analyze the distance between the minimum variance portfolio and the portfolio implemented by the investor for different values of the transaction cost parameter $\beta$ and risk aversion $\gamma$.

```
gammas = [2, 4, 8, 20]
betas = 20*expon.ppf(np.arange(1, 100)/100, scale=1)

transaction_costs = (pd.DataFrame(
    list(product(gammas, betas)),
    columns=["gamma", "beta"]
```

```
)
.assign(
  weights=lambda x: x.apply(lambda y:
    compute_efficient_weight(
      sigma, mu, gamma=y["gamma"], beta=y["beta"]/10000, w_prev=w_mvp),
    axis=1
  ),
  concentration=lambda x: x["weights"].apply(
    lambda x: np.sum(np.abs(x-w_mvp))
  )
)
)
```

The code chunk above computes the optimal weight in the presence of transaction cost for different values of $\beta$ and $\gamma$ but with the same initial allocation, the theoretical optimal minimum variance portfolio. Starting from the initial allocation, the investor chooses their optimal allocation along the efficient frontier to reflect their own risk preferences. If transaction costs were absent, the investor would simply implement the mean-variance efficient allocation. If transaction costs make it costly to rebalance, their optimal portfolio choice reflects a shift toward the efficient portfolio, whereas their current portfolio anchors their investment.

```
rebalancing_plot = (
    ggplot(transaction_costs,
           aes(x="beta", y="concentration",
               color="factor(gamma)", linetype="factor(gamma)")) +
    geom_line() +
    guides(linetype=None) +
    labs(x="Transaction cost parameter", y="Distance from MVP",
        color="Risk aversion",
        title=("Portfolio weights for different risk aversion and "
               "transaction cost"))
)
rebalancing_plot.draw()
```

Figure 18.1 shows rebalancing from the initial portfolio (which we always set to the minimum variance portfolio weights in this example). The higher the transaction costs parameter $\beta$, the smaller is the rebalancing from the initial portfolio. In addition, if risk aversion $\gamma$ increases, the efficient portfolio is closer to the minimum variance portfolio weights such that the investor desires less rebalancing from the initial holdings.

## 18.5 Constrained Optimization

Next, we introduce constraints to the above optimization procedure. Very often, typical constraints such as short-selling restrictions prevent analytical solutions for optimal portfolio weights (short-selling restrictions simply imply that negative weights are not allowed such that we require that $w_i \geq 0 \, \forall i$). However, numerical optimization allows computing the solutions to such constrained problems.

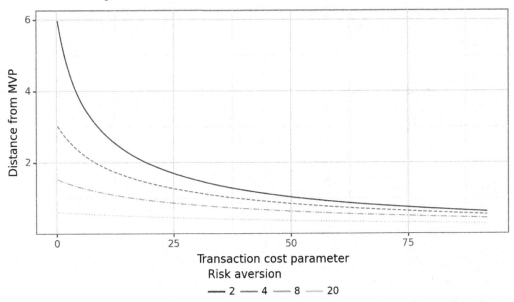

Figure 18.1: The figure shows portfolio weights for different risk aversion and transaction cost. The horizontal axis indicates the distance from the empirical minimum variance portfolio weight, measured by the sum of the absolute deviations of the chosen portfolio from the benchmark.

We rely on the powerful `scipy.optimize` package, which provides a common interface to a number of different optimization routines. In particular, we employ the Sequential Least-Squares Quadratic Programming (SLSQP) algorithm of Kraft (1994) because it is able to handle multiple equality and inequality constraints at the same time and is typically used for problems where the objective function and the constraints are twice continuously differentiable. We have to provide the algorithm with the objective function and its gradient, as well as the constraints and their Jacobian.

We illustrate the use of `minimize()` by replicating the analytical solutions for the minimum variance and efficient portfolio weights from above. Note that the equality constraint for both solutions is given by the requirement that the weights must sum up to one. In addition, we supply a vector of equal weights as an initial value for the algorithm in all applications. We verify that the output is equal to the above solution. Note that `np.allclose()` is a safe way to compare two vectors for pairwise equality. The alternative `==` is sensitive to small differences that may occur due to the representation of floating points on a computer, while `np.allclose()` has a built-in tolerance. It returns `True` if both are equal, which is the case in both applications below.

```
w_initial = np.ones(n_industries)/n_industries

def objective_mvp(w):
    return 0.5*w.T @ sigma @ w

def gradient_mvp(w):
```

```
    return sigma @ w

def equality_constraint(w):
    return np.sum(w)-1

def jacobian_equality(w):
    return np.ones_like(w)

constraints = (
  {"type": "eq", "fun": equality_constraint, "jac": jacobian_equality}
)

options = {
  "tol":1e-20,
  "maxiter": 10000,
  "method":"SLSQP"
}

w_mvp_numerical = minimize(
  x0=w_initial,
  fun=objective_mvp,
  jac=gradient_mvp,
  constraints=constraints,
  tol=options["tol"],
  options={"maxiter": options["maxiter"]},
  method=options["method"]
)

np.allclose(w_mvp, w_mvp_numerical.x, atol=1e-3)

def objective_efficient(w):
    return 2*0.5*w.T @ sigma @ w-(1+mu) @ w

def gradient_efficient(w):
    return 2*sigma @ w-(1+mu)

w_efficient_numerical = minimize(
  x0=w_initial,
  fun=objective_efficient,
  jac=gradient_efficient,
  constraints=constraints,
  tol=options["tol"],
  options={"maxiter": options["maxiter"]},
  method=options["method"]
)

np.allclose(w_efficient, w_efficient_numerical.x, atol = 1e-3)
```

The result above shows that the numerical procedure indeed recovered the optimal weights for a scenario where we already know the analytic solution.

Next, we approach problems where no analytical solutions exist. First, we additionally impose short-sale constraints, which implies $N$ inequality constraints of the form $\omega_i >= 0$. We can implement the short-sale constraints by imposing a vector of lower bounds `lb = rep(0, n_industries)`.

```
w_no_short_sale = minimize(
  x0=w_initial,
  fun=objective_efficient,
  jac=gradient_efficient,
  constraints=constraints,
  bounds=((0, None), )*n_industries,
  tol=options["tol"],
  options={"maxiter": options["maxiter"]},
  method=options["method"]
)

weights_no_short_sale = pd.DataFrame({
  "Industry": industry_returns.columns.tolist(),
  "No short-sale": w_no_short_sale.x
})
weights_no_short_sale.round(3)
```

|   | Industry | No short-sale |
|---|----------|---------------|
| 0 | nodur    | 0.610         |
| 1 | durbl    | 0.000         |
| 2 | manuf    | 0.000         |
| 3 | enrgy    | 0.211         |
| 4 | hitec    | 0.000         |
| 5 | telcm    | 0.000         |
| 6 | shops    | 0.000         |
| 7 | hlth     | 0.179         |
| 8 | utils    | 0.000         |
| 9 | other    | 0.000         |

As expected, the resulting portfolio weights are all positive (up to numerical precision). Typically, the holdings in the presence of short-sale constraints are concentrated among way fewer assets than in the unrestricted case. You can verify that `np.sum(w_no_short_sale.x)` returns 1. In other words, `minimize()` provides the numerical solution to a portfolio choice problem for a mean-variance investor with risk aversion `gamma = 2`, where negative holdings are forbidden.

`minimize()` can also handle more complex problems. As an example, we show how to compute optimal weights, subject to the so-called Regulation-T constraint,[1] which requires that the sum of all absolute portfolio weights is smaller than 1.5, that is $\sum_{i=1}^{N} |\omega_i| \leq 1.5$. The constraint enforces that a maximum of 50 percent of the allocated wealth can be allocated to short positions, thus implying an initial margin requirement of 50 percent. Imposing such a margin requirement reduces portfolio risks because extreme portfolio weights are not attainable anymore. The implementation of Regulation-T rules is numerically interesting because the margin constraints imply a non-linear constraint on the portfolio weights.

---

[1] https://en.wikipedia.org/wiki/Regulation_T

```
reg_t = 1.5

def inequality_constraint(w):
    return reg_t-np.sum(np.abs(w))

def jacobian_inequality(w):
    return -np.sign(w)

def objective_reg_t(w):
    return -w @ (1+mu)+2*0.5*w.T @ sigma @ w

def gradient_reg_t(w):
    return -(1+mu)+2*np.dot(sigma, w)

constraints = (
  {"type": "eq", "fun": equality_constraint, "jac": jacobian_equality},
  {"type": "ineq", "fun": inequality_constraint, "jac": jacobian_inequality}
)

w_reg_t = minimize(
  x0=w_initial,
  fun=objective_reg_t,
  jac=gradient_reg_t,
  constraints=constraints,
  tol=options["tol"],
  options={"maxiter": options["maxiter"]},
  method=options["method"]
)

weights_reg_t = pd.DataFrame({
  "Industry": industry_returns.columns.tolist(),
  "Regulation-T": w_reg_t.x
})
weights_reg_t.round(3)
```

|   | Industry | Regulation-T |
|---|----------|--------------|
| 0 | nodur    | 0.733        |
| 1 | durbl    | -0.000       |
| 2 | manuf    | -0.144       |
| 3 | enrgy    | 0.264        |
| 4 | hitec    | 0.000        |
| 5 | telcm    | -0.000       |
| 6 | shops    | 0.031        |
| 7 | hlth     | 0.222        |
| 8 | utils    | -0.000       |
| 9 | other    | -0.106       |

Figure 18.2 shows the optimal allocation weights across all `python len(industry_returns.columns)` industries for the four different strategies considered

so far: minimum variance, efficient portfolio with $\gamma = 2$, efficient portfolio with short-sale constraints, and the Regulation-T constrained portfolio.

```python
weights = (weights_mvp
  .merge(weights_efficient)
  .merge(weights_no_short_sale)
  .merge(weights_reg_t)
  .melt(id_vars="Industry", var_name="Strategy", value_name="weights")
)

weights_plot = (
  ggplot(weights,
         aes(x="Industry", y="weights", fill="Strategy")) +
  geom_bar(stat="identity", position="dodge", width=0.7) +
  coord_flip() +
  labs(y="Allocation weight", fill="",
       title="Optimal allocations for different strategies") +
  scale_y_continuous(labels=percent_format())
)
weights_plot.draw()
```

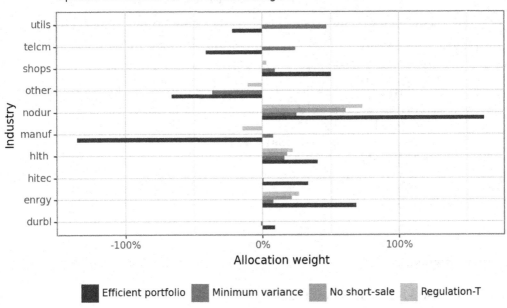

Figure 18.2: The figure shows optimal allocation weights for the ten industry portfolios and the four different allocation strategies.

The results clearly indicate the effect of imposing additional constraints: the extreme holdings the investor implements if they follow the (theoretically optimal) efficient portfolio vanish under, e.g., the Regulation-T constraint. You may wonder why an investor would deviate from what is theoretically the optimal portfolio by imposing potentially arbitrary constraints. The short answer is: the *efficient portfolio* is only efficient if the true parameters of the data-generating process correspond to the estimated parameters $\hat{\Sigma}$ and $\hat{\mu}$. Estimation uncertainty may thus lead to inefficient allocations. By imposing restrictions, we implicitly shrink the set of possible weights and prevent extreme allocations, which could result from *error-maximization* due to estimation uncertainty (Jagannathan and Ma, 2003).

Before we move on, we want to propose a final allocation strategy, which reflects a somewhat more realistic structure of transaction costs instead of the quadratic specification used above. The function below computes efficient portfolio weights while adjusting for transaction costs of the form $\beta \sum_{i=1}^{N} |(\omega_{i,t+1} - \omega_{i,t^+})|$. No closed-form solution exists, and we rely on non-linear optimization procedures.

```python
def compute_efficient_weight_L1_TC(mu, sigma, gamma, beta, initial_weights):
    """Compute efficient portfolio weights with L1 constraint."""

    def objective(w):
        return (gamma*0.5*w.T @ sigma @ w-(1+mu) @ w
                +(beta/10000)/2*np.sum(np.abs(w-initial_weights)))

    def gradient(w):
        return (-mu+gamma*sigma @ w
                +(beta/10000)*0.5*np.sign(w-initial_weights))

    constraints = (
        {"type": "eq", "fun": equality_constraint, "jac": jacobian_equality}
    )

    result = minimize(
        x0=initial_weights,
        fun=objective,
        jac=gradient,
        constraints=constraints,
        tol=options["tol"],
        options={"maxiter": options["maxiter"]},
        method=options["method"]
    )

    return result.x
```

## 18.6   Out-of-Sample Backtesting

For the sake of simplicity, we committed one fundamental error in computing portfolio weights above: we used the full sample of the data to determine the optimal allocation (Arnott et al., 2019). To implement this strategy at the beginning of the 2000s, you will need to know how the returns will evolve until 2021. While interesting from a methodological point of view, we cannot evaluate the performance of the portfolios in a reasonable out-of-sample fashion. We do so next in a backtesting application for three strategies. For the backtest, we recompute optimal weights just based on past available data.

The few lines below define the general setup. We consider 120 periods from the past to update the parameter estimates before recomputing portfolio weights. Then, we update portfolio weights, which is costly and affects the performance. The portfolio weights determine the portfolio return. A period later, the current portfolio weights have changed and form the foundation for transaction costs incurred in the next period. We consider three different competing strategies: the mean-variance efficient portfolio, the mean-variance efficient portfolio with ex-ante adjustment for transaction costs, and the naive portfolio, which allocates wealth equally across the different assets.

```
window_length = 120
periods = industry_returns.shape[0]-window_length

beta = 50
gamma = 2

performance_values = np.empty((periods, 3))
performance_values[:] = np.nan
performance_values = {
  "MV (TC)": performance_values.copy(),
  "Naive": performance_values.copy(),
  "MV": performance_values.copy()
}

n_industries = industry_returns.shape[1]
w_prev_1 = w_prev_2 = w_prev_3 = np.ones(n_industries)/n_industries
```

We also define two helper functions: One to adjust the weights due to returns and one for performance evaluation, where we compute realized returns net of transaction costs.

```
def adjust_weights(w, next_return):
    w_prev = 1+w*next_return
    return np.array(w_prev/np.sum(np.array(w_prev)))

def evaluate_performance(w, w_previous, next_return, beta=50):
    """Calculate portfolio evaluation measures."""

    raw_return = np.dot(next_return, w)
    turnover = np.sum(np.abs(w-w_previous))
    net_return = raw_return-beta/10000*turnover

    return np.array([raw_return, turnover, net_return])
```

The following code chunk performs a rolling-window estimation, which we implement in a loop. In each period, the estimation window contains the returns available up to the current period. Note that we use the sample variance-covariance matrix and ignore the estimation of $\hat{\mu}$ entirely, but you might use more advanced estimators in practice.

```
for p in range(periods):
    returns_window = industry_returns.iloc[p:(p+window_length-1), :]
    next_return = industry_returns.iloc[p+window_length, :]

    sigma_window = np.array(returns_window.cov())
    mu = 0*np.array(returns_window.mean())

    # Transaction-cost adjusted portfolio
```

```
    w_1 = compute_efficient_weight_L1_TC(
      mu=mu, sigma=sigma_window,
      beta=beta,
      gamma=gamma,
      initial_weights=w_prev_1
    )

    performance_values["MV (TC)"][p, :] = evaluate_performance(
      w_1, w_prev_1, next_return, beta=beta
    )
    w_prev_1 = adjust_weights(w_1, next_return)

    # Naive portfolio
    w_2 = np.ones(n_industries)/n_industries
    performance_values["Naive"][p, :] = evaluate_performance(
      w_2, w_prev_2, next_return
    )
    w_prev_2 = adjust_weights(w_2, next_return)

    # Mean-variance efficient portfolio (w/o transaction costs)
    w_3 = compute_efficient_weight(sigma=sigma_window, mu=mu, gamma=gamma)
    performance_values["MV"][p, :] = evaluate_performance(
      w_3, w_prev_3, next_return
    )
    w_prev_3 = adjust_weights(w_3, next_return)
```

Finally, we get to the evaluation of the portfolio strategies *net-of-transaction costs*. Note that we compute annualized returns and standard deviations.

```
performance = pd.DataFrame()
for i in enumerate(performance_values.keys()):
    tmp_data = pd.DataFrame(
      performance_values[i[1]],
      columns=["raw_return", "turnover", "net_return"]
    )
    tmp_data["strategy"] = i[1]
    performance = pd.concat([performance, tmp_data], axis=0)

length_year = 12

performance_table = (performance
  .groupby("strategy")
  .aggregate(
    mean=("net_return", lambda x: length_year*100*x.mean()),
    sd=("net_return", lambda x: np.sqrt(length_year)*100*x.std()),
    sharpe_ratio=("net_return", lambda x: (
      (length_year*100*x.mean())/(np.sqrt(length_year)*100*x.std())
        if x.mean() > 0 else np.nan)
    ),
    turnover=("turnover", lambda x: 100*x.mean())
  )
```

```
    .reset_index()
)
performance_table.round(3)
```

| | strategy | mean | sd | sharpe_ratio | turnover |
|---|---|---|---|---|---|
| 0 | MV | -0.896 | 12.571 | NaN | 211.095 |
| 1 | MV (TC) | 11.853 | 15.189 | 0.780 | 0.000 |
| 2 | Naive | 11.832 | 15.191 | 0.779 | 0.234 |

The results clearly speak against mean-variance optimization. Turnover is huge when the investor only considers their portfolio's expected return and variance. Effectively, the mean-variance portfolio generates a *negative* annualized return after adjusting for transaction costs. At the same time, the naive portfolio turns out to perform very well. In fact, the performance gains of the transaction-cost adjusted mean-variance portfolio are small. The out-of-sample Sharpe ratio is slightly higher than for the naive portfolio. Note the extreme effect of turnover penalization on turnover: *MV (TC)* effectively resembles a buy-and-hold strategy which only updates the portfolio once the estimated parameters $\hat{\mu}_t$ and $\hat{\Sigma}_t$ indicate that the current allocation is too far away from the optimal theoretical portfolio.

## 18.7   Exercises

1. Consider the portfolio choice problem for transaction-cost adjusted certainty equivalent maximization with risk aversion parameter $\gamma$

$$\omega_{t+1}^* = \arg\max_{\omega \in \mathbb{R}^N, \iota'\omega=1} \omega'\mu - \nu_t(\omega, \beta) - \frac{\gamma}{2}\omega'\Sigma\omega \qquad (18.10)$$

   where $\Sigma$ and $\mu$ are (estimators of) the variance-covariance matrix of the returns and the vector of expected returns. Assume for now that transaction costs are quadratic in rebalancing *and* proportional to stock illiquidity such that

$$\nu_t(\omega, B) = \frac{\beta}{2}(\omega - \omega_{t^+})' B(\omega - \omega_{t^+}) \qquad (18.11)$$

   where $B = \text{diag}(ill_1, \dots, ill_N)$ is a diagonal matrix, where $ill_1, \dots, ill_N$. Derive a closed-form solution for the mean-variance efficient portfolio $\omega_{t+1}^*$ based on the transaction cost specification above. Discuss the effect of illiquidity $ill_i$ on the individual portfolio weights relative to an investor that myopically ignores transaction costs in their decision.

2. Use the solution from the previous exercise to update the function `compute_efficient_weight()` such that you can compute optimal weights conditional on a matrix $B$ with illiquidity measures.

3. Illustrate the evolution of the *optimal* weights from the naive portfolio to the efficient portfolio in the mean-standard deviation diagram.

4. Is it always optimal to choose the same $\beta$ in the optimization problem than the value used in evaluating the portfolio performance? In other words, can it be optimal to choose theoretically sub-optimal portfolios based on transaction cost

considerations that do not reflect the actual incurred costs? Evaluate the out-of-sample Sharpe ratio after transaction costs for a range of different values of imposed $\beta$ values.

# A

## *Colophon*

In this appendix chapter, we provide details on the package versions used in this edition. The book was built with Python (Python Software Foundation, 2023) version 3.10.11 and the following packages:

|    | Package | Version |
| --- | --- | --- |
| 0 | anyio | 4.0.0 |
| 1 | appdirs | 1.4.4 |
| 2 | appnope | 0.1.3 |
| 3 | argon2-cffi | 23.1.0 |
| 4 | argon2-cffi-bindings | 21.2.0 |
| 5 | arrow | 1.3.0 |
| 6 | astor | 0.8.1 |
| 7 | asttokens | 2.4.0 |
| 8 | async-lru | 2.0.4 |
| 9 | attrs | 23.1.0 |
| 10 | Babel | 2.13.0 |
| 11 | backcall | 0.2.0 |
| 12 | beautifulsoup4 | 4.12.2 |
| 13 | bleach | 6.1.0 |
| 14 | certifi | 2023.7.22 |
| 15 | cffi | 1.16.0 |
| 16 | charset-normalizer | 3.3.0 |
| 17 | click | 8.1.7 |
| 18 | comm | 0.1.4 |
| 19 | contourpy | 1.1.1 |
| 20 | cycler | 0.12.1 |
| 21 | Cython | 3.0.3 |
| 22 | debugpy | 1.8.0 |
| 23 | decorator | 5.1.1 |
| 24 | defusedxml | 0.7.1 |
| 25 | et-xmlfile | 1.1.0 |
| 26 | exceptiongroup | 1.1.3 |
| 27 | executing | 2.0.0 |
| 28 | fastjsonschema | 2.18.1 |
| 29 | fonttools | 4.43.1 |
| 30 | formulaic | 0.6.6 |
| 31 | fqdn | 1.5.1 |
| 32 | frozendict | 2.3.8 |
| 33 | html5lib | 1.1 |
| 34 | httpimport | 1.3.1 |

|    | Package                      | Version   |
|----|------------------------------|-----------|
| 35 | idna                         | 3.4       |
| 36 | importlib-metadata           | 6.8.0     |
| 37 | interface-meta               | 1.3.0     |
| 38 | ipykernel                    | 6.25.2    |
| 39 | ipython                      | 8.16.1    |
| 40 | ipython-genutils             | 0.2.0     |
| 41 | ipywidgets                   | 8.1.1     |
| 42 | isoduration                  | 20.11.0   |
| 43 | jedi                         | 0.18.2    |
| 44 | Jinja2                       | 3.1.2     |
| 45 | joblib                       | 1.3.2     |
| 46 | json5                        | 0.9.14    |
| 47 | jsonpointer                  | 2.4       |
| 48 | jsonschema                   | 4.19.1    |
| 49 | jsonschema-specifications    | 2023.7.1  |
| 50 | jupyter                      | 1.0.0     |
| 51 | jupyter-cache                | 0.6.1     |
| 52 | jupyter-console              | 6.6.3     |
| 53 | jupyter-events               | 0.7.0     |
| 54 | jupyter-lsp                  | 2.2.0     |
| 55 | jupyter_client               | 8.3.1     |
| 56 | jupyter_core                 | 5.4.0     |
| 57 | jupyter_server               | 2.7.3     |
| 58 | jupyter_server_terminals     | 0.4.4     |
| 59 | jupyterlab                   | 4.0.6     |
| 60 | jupyterlab-pygments          | 0.2.2     |
| 61 | jupyterlab-widgets           | 3.0.9     |
| 62 | jupyterlab_server            | 2.25.0    |
| 63 | kiwisolver                   | 1.4.5     |
| 64 | linearmodels                 | 5.3       |
| 65 | lxml                         | 4.9.3     |
| 66 | MarkupSafe                   | 2.1.3     |
| 67 | matplotlib                   | 3.8.0     |
| 68 | matplotlib-inline            | 0.1.6     |
| 69 | mistune                      | 3.0.2     |
| 70 | mizani                       | 0.9.3     |
| 71 | multitasking                 | 0.0.11    |
| 72 | mypy-extensions              | 1.0.0     |
| 73 | nbclient                     | 0.7.4     |
| 74 | nbconvert                    | 7.9.2     |
| 75 | nbformat                     | 5.9.2     |
| 76 | nest-asyncio                 | 1.5.8     |
| 77 | notebook                     | 7.0.4     |
| 78 | notebook_shim                | 0.2.3     |
| 79 | numpy                        | 1.26.0    |
| 80 | openpyxl                     | 3.1.2     |
| 81 | overrides                    | 7.4.0     |
| 82 | packaging                    | 23.2      |
| 83 | pandas                       | 2.1.1     |

|     | Package | Version |
| --- | --- | --- |
| 84 | pandas-datareader | 0.10.0 |
| 85 | pandocfilters | 1.5.0 |
| 86 | parso | 0.8.3 |
| 87 | patsy | 0.5.3 |
| 88 | peewee | 3.16.3 |
| 89 | pexpect | 4.8.0 |
| 90 | pickleshare | 0.7.5 |
| 91 | Pillow | 10.0.1 |
| 92 | platformdirs | 3.11.0 |
| 93 | plotnine | 0.12.3 |
| 94 | prometheus-client | 0.17.1 |
| 95 | prompt-toolkit | 3.0.39 |
| 96 | psutil | 5.9.5 |
| 97 | psycopg2-binary | 2.9.9 |
| 98 | ptyprocess | 0.7.0 |
| 99 | pure-eval | 0.2.2 |
| 100 | pycparser | 2.21 |
| 101 | Pygments | 2.16.1 |
| 102 | pyhdfe | 0.2.0 |
| 103 | pyparsing | 3.1.1 |
| 104 | python-dateutil | 2.8.2 |
| 105 | python-dotenv | 1.0.0 |
| 106 | python-json-logger | 2.0.7 |
| 107 | pytz | 2023.3.post1 |
| 108 | PyYAML | 6.0.1 |
| 109 | pyzmq | 25.1.1 |
| 110 | qtconsole | 5.4.4 |
| 111 | QtPy | 2.4.0 |
| 112 | referencing | 0.30.2 |
| 113 | regtabletotext | 0.0.12 |
| 114 | requests | 2.31.0 |
| 115 | rfc3339-validator | 0.1.4 |
| 116 | rfc3986-validator | 0.1.1 |
| 117 | rpds-py | 0.10.4 |
| 118 | scikit-learn | 1.3.1 |
| 119 | scipy | 1.11.3 |
| 120 | Send2Trash | 1.8.2 |
| 121 | setuptools-scm | 7.1.0 |
| 122 | six | 1.16.0 |
| 123 | sniffio | 1.3.0 |
| 124 | soupsieve | 2.5 |
| 125 | SQLAlchemy | 2.0.21 |
| 126 | stack-data | 0.6.3 |
| 127 | statsmodels | 0.14.0 |
| 128 | tabulate | 0.9.0 |
| 129 | terminado | 0.17.1 |
| 130 | threadpoolctl | 3.2.0 |
| 131 | tinycss2 | 1.2.1 |
| 132 | tomli | 2.0.1 |

| | Package | Version |
|---|---|---|
| 133 | tornado | 6.3.3 |
| 134 | traitlets | 5.11.2 |
| 135 | types-python-dateutil | 2.8.19.14 |
| 136 | typing_extensions | 4.8.0 |
| 137 | tzdata | 2023.3 |
| 138 | uri-template | 1.3.0 |
| 139 | urllib3 | 2.0.6 |
| 140 | wcwidth | 0.2.8 |
| 141 | webcolors | 1.13 |
| 142 | webencodings | 0.5.1 |
| 143 | websocket-client | 1.6.4 |
| 144 | widgetsnbextension | 4.0.9 |
| 145 | wrapt | 1.15.0 |
| 146 | yfinance | 0.2.31 |
| 147 | zipp | 3.17.0 |

# B

## Proofs

In this appendix chapter, we collect the proofs that we refer to throughout the book.

### B.1 Optimal Portfolio Choice

#### B.1.1 Minimum variance portfolio

The minimum variance portfolio weights are given by the solution to

$$\omega_{\text{mvp}} = \arg \min \omega' \Sigma \omega \text{ s.t. } \iota' \omega = 1,$$

where $\iota$ is an $(N \times 1)$ vector of ones. The Lagrangian reads

$$\mathcal{L}(\omega) = \omega' \Sigma \omega - \lambda(\omega' \iota - 1).$$

We can solve the first-order conditions of the Lagrangian equation:

$$\frac{\partial \mathcal{L}(\omega)}{\partial \omega} = 0 \Leftrightarrow 2\Sigma \omega = \lambda \iota \Rightarrow \omega = \frac{\lambda}{2} \Sigma^{-1} \iota$$

Next, the constraint that weights have to sum up to one delivers: $1 = \iota' \omega = \frac{\lambda}{2} \iota' \Sigma^{-1} \iota \Rightarrow \lambda = \frac{2}{\iota' \Sigma^{-1} \iota}$. Finally, plug-in the derived value of $\lambda$ to get

$$\omega_{\text{mvp}} = \frac{\Sigma^{-1} \iota}{\iota' \Sigma^{-1} \iota}.$$

#### B.1.2 Efficient portfolio

Consider an investor who aims to achieve minimum variance *given a desired expected return* $\bar{\mu}$, that is:

$$\omega_{\text{eff}}(\bar{\mu}) = \arg \min \omega' \Sigma \omega \text{ s.t. } \iota' \omega = 1 \text{ and } \omega' \mu \geq \bar{\mu}.$$

The Lagrangian reads

$$\mathcal{L}(\omega) = \omega' \Sigma \omega - \lambda(\omega' \iota - 1) - \tilde{\lambda}(\omega' \mu - \bar{\mu}).$$

We can solve the first-order conditions to get

$$2\Sigma \omega = \lambda \iota + \tilde{\lambda} \mu$$

$$\Rightarrow \omega = \frac{\lambda}{2} \Sigma^{-1} \iota + \frac{\tilde{\lambda}}{2} \Sigma^{-1} \mu.$$

Next, the two constraints ($w'\iota = 1$ and $\omega'\mu \geq \bar{\mu}$) imply

$$1 = \iota'\omega = \frac{\lambda}{2}\underbrace{\iota'\Sigma^{-1}\iota}_{C} + \frac{\tilde{\lambda}}{2}\underbrace{\iota'\Sigma^{-1}\mu}_{D}$$

$$\Rightarrow \lambda = \frac{2 - \tilde{\lambda}D}{C}$$

$$\bar{\mu} = \mu'\omega = \frac{\lambda}{2}\underbrace{\mu'\Sigma^{-1}\iota}_{D} + \frac{\tilde{\lambda}}{2}\underbrace{\mu'\Sigma^{-1}\mu}_{E} = \frac{1}{2}\left(\frac{2 - \tilde{\lambda}D}{C}\right)D + \frac{\tilde{\lambda}}{2}E$$

$$= \frac{D}{C} + \frac{\tilde{\lambda}}{2}\left(E - \frac{D^2}{C}\right)$$

$$\Rightarrow \tilde{\lambda} = 2\frac{\bar{\mu} - D/C}{E - D^2/C}.$$

As a result, the efficient portfolio weight takes the form (for $\bar{\mu} \geq D/C = \mu'\omega_{\mathrm{mvp}}$)

$$\omega_{\mathrm{eff}}(\bar{\mu}) = \omega_{\mathrm{mvp}} + \frac{\tilde{\lambda}}{2}\left(\Sigma^{-1}\mu - \frac{D}{C}\Sigma^{-1}\iota\right).$$

Thus, the efficient portfolio allocates wealth in the minimum variance portfolio $\omega_{\mathrm{mvp}}$ and a levered (self-financing) portfolio to increase the expected return.

Note that the portfolio weights sum up to one as

$$\iota'\left(\Sigma^{-1}\mu - \frac{D}{C}\Sigma^{-1}\iota\right) = D - D = 0 \text{ so } \iota'\omega_{\mathrm{eff}} = \iota'\omega_{\mathrm{mvp}} = 1.$$

Finally, the expected return of the efficient portfolio is

$$\mu'\omega_{\mathrm{eff}} = \frac{D}{C} + \bar{\mu} - \frac{D}{C} = \bar{\mu}.$$

### B.1.3   Equivalence between certainty equivalent maximization and minimum variance optimization

We argue that an investor with a quadratic utility function with certainty equivalent

$$\max_{\omega} CE(\omega) = \omega'\mu - \frac{\gamma}{2}\omega'\Sigma\omega \text{ s.t. } \iota'\omega = 1$$

faces an equivalent optimization problem to a framework where portfolio weights are chosen with the aim to minimize volatility given a pre-specified level or expected returns

$$\min_{\omega} \omega'\Sigma\omega \text{ s.t. } \omega'\mu = \bar{\mu} \text{ and } \iota'\omega = 1.$$

Note the difference: In the first case, the investor has a (known) risk aversion $\gamma$ which determines their optimal balance between risk ($\omega'\Sigma\omega$) and return ($\mu'\omega$). In the second case, the investor has a target return they want to achieve while minimizing the volatility. Intuitively, both approaches are closely connected if we consider that the risk aversion $\gamma$ determines the desirable return $\bar{\mu}$. More risk-averse investors (higher $\gamma$) will chose a lower target return to keep their volatility level down. The efficient frontier then spans all possible portfolios depending on the risk aversion $\gamma$, starting from the minimum variance portfolio ($\gamma = \infty$).

To proof this equivalence, consider first the optimal portfolio weights for a certainty equivalent maximizing investor. The first-order condition reads

$$\mu - \lambda \iota = \gamma \Sigma \omega$$

$$\Leftrightarrow \omega = \frac{1}{\gamma} \Sigma^{-1} (\mu - \lambda \iota)$$

Next, we make use of the constraint $\iota' \omega = 1$.

$$\iota' \omega = 1 = \frac{1}{\gamma} \left( \iota' \Sigma^{-1} \mu - \lambda \iota' \Sigma^{-1} \iota \right)$$

$$\Rightarrow \lambda = \frac{1}{\iota' \Sigma^{-1} \iota} \left( \iota' \Sigma^{-1} \mu - \gamma \right).$$

Plugging in the value of $\lambda$ reveals the desired portfolio for an investor with risk aversion $\gamma$.

$$\omega = \frac{1}{\gamma} \Sigma^{-1} \left( \mu - \frac{1}{\iota' \Sigma^{-1} \iota} \left( \iota' \Sigma^{-1} \mu - \gamma \right) \right)$$

$$\Rightarrow \omega = \frac{\Sigma^{-1} \iota}{\iota' \Sigma^{-1} \iota} + \frac{1}{\gamma} \left( \Sigma^{-1} - \frac{\Sigma^{-1} \iota}{\iota' \Sigma^{-1} \iota} \iota' \Sigma^{-1} \right) \mu$$

$$= \omega_{\text{mvp}} + \frac{1}{\gamma} \left( \Sigma^{-1} \mu - \frac{\iota' \Sigma^{-1} \mu}{\iota' \Sigma^{-1} \iota} \Sigma^{-1} \iota \right).$$

The resulting weights correspond to the efficient portfolio with desired return $\bar{r}$ such that (in the notation of book)

$$\frac{1}{\gamma} = \frac{\tilde{\lambda}}{2} = \frac{\bar{\mu} - D/C}{E - D^2/C}$$

which implies that the desired return is just

$$\bar{\mu} = \frac{D}{C} + \frac{1}{\gamma} \left( E - D^2/C \right)$$

which is $\frac{D}{C} = \mu' \omega_{\text{mvp}}$ for $\gamma \to \infty$ as expected. For instance, letting $\gamma \to \infty$ implies $\bar{\mu} = \frac{D}{C} = \omega'_{\text{mvp}} \mu$.

# C

## WRDS Dummy Data

In this appendix chapter, we alleviate the constraints of readers who don't have access to WRDS and hence cannot run the code that we provide. We show how to create a dummy database that contains the WRDS tables and corresponding columns such that all code chunks in this book can be executed with this dummy database. We do not create dummy data for tables of macroeconomic variables because they can be freely downloaded from the original sources; check out Chapter 3.

We deliberately use the dummy label because the data is not meaningful in the sense that it allows readers to actually replicate the results of the book. For legal reasons, the data does not contain any samples of the original data. We merely generate random numbers for all columns of the tables that we use throughout the books.

To generate the dummy database, we use the following packages:

```python
import pandas as pd
import numpy as np
import sqlite3
import string
```

Let us initialize a SQLite database (`tidy_finance_python.sqlite`) or connect to your existing one. Be careful, if you already downloaded the data from WRDS, then the code in this chapter will overwrite your data!

```python
tidy_finance = sqlite3.connect(database="data/tidy_finance_python.sqlite")
```

Since we draw random numbers for most of the columns, we also define a seed to ensure that the generated numbers are replicable. We also initialize vectors of dates of different frequencies over ten years that we then use to create yearly, monthly, and daily data, respectively.

```python
np.random.seed(1234)

start_date = pd.Timestamp("2003-01-01")
end_date = pd.Timestamp("2022-12-31")

dummy_years = np.arange(start_date.year, end_date.year+1, 1)
dummy_months = pd.date_range(start_date, end_date, freq="M")
dummy_days = pd.date_range(start_date, end_date, freq="D")
```

## C.1   Create Stock Dummy Data

Let us start with the core data used throughout the book: stock and firm characteristics. We first generate a table with a cross-section of stock identifiers with unique `permno` and `gvkey` values, as well as associated `exchcd`, `exchange`, `industry`, and `siccd` values. The generated data is based on the characteristics of stocks in the `crsp_monthly` table of the original database, ensuring that the generated stocks roughly reflect the distribution of industries and exchanges in the original data, but the identifiers and corresponding exchanges or industries do not reflect actual firms. Similarly, the `permno-gvkey` combinations are purely nonsensical and should not be used together with actual CRSP or Compustat data.

```python
number_of_stocks = 100

industries = pd.DataFrame({
  "industry": ["Agriculture", "Construction", "Finance",
               "Manufacturing", "Mining", "Public", "Retail",
               "Services", "Transportation", "Utilities", "Wholesale"],
  "n": [81, 287, 4682, 8584, 1287, 1974, 1571, 4277, 1249, 457, 904],
  "prob": [0.00319, 0.0113, 0.185, 0.339, 0.0508, 0.0779,
           0.0620, 0.169, 0.0493, 0.0180, 0.03451]
})

exchanges = pd.DataFrame({
  "exchange": ["AMEX", "NASDAQ", "NYSE"],
  "n": [2893, 17236, 5553],
  "prob": [0.113, 0.671, 0.216]
})

stock_identifiers_list = []
for x in range(1, number_of_stocks+1):
  exchange = np.random.choice(exchanges["exchange"], p=exchanges["prob"])
  industry = np.random.choice(industries["industry"], p=industries["prob"])

  exchcd_mapping = {
    "NYSE": np.random.choice([1, 31]),
    "AMEX": np.random.choice([2, 32]),
    "NASDAQ": np.random.choice([3, 33])
  }

  siccd_mapping = {
    "Agriculture": np.random.randint(1, 1000),
    "Mining": np.random.randint(1000, 1500),
    "Construction": np.random.randint(1500, 1800),
    "Manufacturing": np.random.randint(1800, 4000),
    "Transportation": np.random.randint(4000, 4900),
    "Utilities": np.random.randint(4900, 5000),
    "Wholesale": np.random.randint(5000, 5200),
    "Retail": np.random.randint(5200, 6000),
    "Finance": np.random.randint(6000, 6800),
```

```
      "Services": np.random.randint(7000, 9000),
      "Public": np.random.randint(9000, 10000)
  }

  stock_identifiers_list.append({
      "permno": x,
      "gvkey": str(x+10000),
      "exchange": exchange,
      "industry": industry,
      "exchcd": exchcd_mapping[exchange],
      "siccd": siccd_mapping[industry]
  })

stock_identifiers = pd.DataFrame(stock_identifiers_list)
```

Next, we construct three panels of stock data with varying frequencies: yearly, monthly, and daily. We begin by creating the `stock_panel_yearly` panel. To achieve this, we combine the `stock_identifiers` table with a new table containing the variable `year` from `dummy_years`. The `expand_grid()` function ensures that we get all possible combinations of the two tables. After combining, we select only the `gvkey` and `year` columns for our final yearly panel.

Next, we construct the `stock_panel_monthly` panel. Similar to the yearly panel, we use the `expand_grid()` function to combine `stock_identifiers` with a new table that has the `month` variable from `dummy_months`. After merging, we select the columns `permno`, `gvkey`, `month`, `siccd`, `industry`, `exchcd`, and `exchange` to form our monthly panel.

Lastly, we create the `stock_panel_daily` panel. We combine `stock_identifiers` with a table containing the `date` variable from `dummy_days`. After merging, we retain only the `permno` and `date` columns for our daily panel.

```
stock_panel_yearly = pd.DataFrame({
  "gvkey": np.tile(stock_identifiers["gvkey"], len(dummy_years)),
  "year": np.repeat(dummy_years, len(stock_identifiers))
})

stock_panel_monthly = pd.DataFrame({
  "permno": np.tile(stock_identifiers["permno"], len(dummy_months)),
  "gvkey": np.tile(stock_identifiers["gvkey"], len(dummy_months)),
  "month": np.repeat(dummy_months, len(stock_identifiers)),
  "siccd": np.tile(stock_identifiers["siccd"], len(dummy_months)),
  "industry": np.tile(stock_identifiers["industry"], len(dummy_months)),
  "exchcd": np.tile(stock_identifiers["exchcd"], len(dummy_months)),
  "exchange": np.tile(stock_identifiers["exchange"], len(dummy_months))
})

stock_panel_daily = pd.DataFrame({
  "permno": np.tile(stock_identifiers["permno"], len(dummy_days)),
  "date": np.repeat(dummy_days, len(stock_identifiers))
})
```

## C.1.1  Dummy `beta` table

We then proceed to create dummy beta values for our `stock_panel_monthly` table. We generate monthly beta values `beta_monthly` using the `rnorm()` function with a mean and standard deviation of 1. For daily beta values `beta_daily`, we take the dummy monthly beta and add a small random noise to it. This noise is generated again using the `rnorm()` function, but this time we divide the random values by 100 to ensure they are small deviations from the monthly beta.

```python
beta_dummy = (stock_panel_monthly
  .assign(
    beta_monthly=np.random.normal(
      loc=1, scale=1, size=len(stock_panel_monthly)
    ),
    beta_daily=lambda x: (
      x["beta_monthly"]+np.random.normal(scale=0.01, size=len(x))
    )
  )
)

(beta_dummy
  .to_sql(name="beta",
          con=tidy_finance,
          if_exists="replace",
          index = False)
)
```

## C.1.2  Dummy `compustat` table

To create dummy firm characteristics, we take all columns from the `compustat` table and create random numbers between 0 and 1. For simplicity, we set the `datadate` for each firm-year observation to the last day of the year, although it is empirically not the case.

```python
relevant_columns = [
  "seq", "ceq", "at", "lt", "txditc", "txdb", "itcb",
  "pstkrv", "pstkl", "pstk", "capx", "oancf", "sale",
  "cogs", "xint", "xsga", "be", "op", "at_lag", "inv"
]

commands = {
  col: np.random.rand(len(stock_panel_yearly)) for col in relevant_columns
}

compustat_dummy = (
  stock_panel_yearly
  .assign(
    datadate=lambda x: pd.to_datetime(x["year"].astype(str)+"-12-31")
  )
  .assign(**commands)
)

(compustat_dummy
```

```
        .to_sql(name="compustat",
                con=tidy_finance,
                if_exists="replace",
                index=False)
)
```

### C.1.3  Dummy `crsp_monthly` table

The `crsp_monthly` table only lacks a few more columns compared to `stock_panel_monthly`: the returns `ret` drawn from a normal distribution, the excess returns `ret_excess` with small deviations from the returns, the shares outstanding `shrout` and the last price per month `altprc` both drawn from uniform distributions, and the market capitalization `mktcap` as the product of `shrout` and `altprc`.

```
crsp_monthly_dummy = (stock_panel_monthly
  .assign(
    date=lambda x: x["month"]+pd.offsets.MonthEnd(-1),
    ret=lambda x: np.fmax(np.random.normal(size=len(x)), -1),
    ret_excess=lambda x: (
      np.fmax(x["ret"]-np.random.uniform(0, 0.0025, len(x)), -1)
    ),
    shrout=1000*np.random.uniform(1, 50, len(stock_panel_monthly)),
    altprc=np.random.uniform(0, 1000, len(stock_panel_monthly)))
  .assign(mktcap=lambda x: x["shrout"]*x["altprc"])
  .sort_values(by=["permno", "month"])
  .assign(
    mktcap_lag=lambda x: (x.groupby("permno")["mktcap"].shift(1))
  )
  .reset_index(drop=True)
)

(crsp_monthly_dummy
  .to_sql(name="crsp_monthly",
          con=tidy_finance,
          if_exists="replace",
          index=False)
)
```

### C.1.4  Dummy `crsp_daily` table

The `crsp_daily` table only contains a `month` column and the daily excess returns `ret_excess` as additional columns to `stock_panel_daily`.

```
crsp_daily_dummy = (stock_panel_daily
  .assign(
    month=lambda x: x["date"]-pd.offsets.MonthBegin(1),
    ret_excess=lambda x: np.fmax(np.random.normal(size=len(x)), -1)
  )
  .reset_index(drop=True)
)
```

```
(crsp_daily_dummy
  .to_sql(name="crsp_daily",
          con=tidy_finance,
          if_exists="replace",
          index=False)
)
```

## C.2   Create Bond Dummy Data

Lastly, we move to the bond data that we use in our books.

### C.2.1   Dummy `fisd` data

To create dummy data with the structure of Mergent FISD, we calculate the empirical probabilities of actual bonds for several variables: `maturity`, `offering_amt`, `interest_frequency`, `coupon`, and `sic_code`. We use these probabilities to sample a small cross-section of bonds with completely made up `complete_cusip`, `issue_id`, and `issuer_id`.

```
number_of_bonds = 100

def generate_cusip():
  """Generate cusip."""

  characters = list(string.ascii_uppercase+string.digits)  # Convert to list
  cusip = ("".join(np.random.choice(characters, size=12))).upper()

  return cusip

fisd_dummy = (pd.DataFrame({
    "complete_cusip": [generate_cusip() for _ in range(number_of_bonds)]
  })
  .assign(
    maturity=lambda x: np.random.choice(dummy_days, len(x), replace=True),
    offering_amt=lambda x: np.random.choice(
      np.arange(1, 101)*100000, len(x), replace=True
    )
  )
  .assign(
    offering_date=lambda x: (
      x["maturity"]-pd.to_timedelta(
        np.random.choice(np.arange(1, 26)*365, len(x), replace=True),
        unit="D"
      )
    )
  )
  .assign(
```

```
          dated_date=lambda x: (
            x["offering_date"]-pd.to_timedelta(
              np.random.choice(np.arange(-10, 11), len(x), replace=True),
              unit="D"
            )
          ),
          interest_frequency=lambda x: np.random.choice(
            [0, 1, 2, 4, 12], len(x), replace=True
          ),
          coupon=lambda x: np.random.choice(
            np.arange(0, 2.1, 0.1), len(x), replace=True
          )
        )
        .assign(
          last_interest_date=lambda x: (
            x[["maturity", "offering_date", "dated_date"]].max(axis=1)
          ),
          issue_id=lambda x: x.index+1,
          issuer_id=lambda x: np.random.choice(
            np.arange(1, 251), len(x), replace=True
          ),
          sic_code=lambda x: (np.random.choice(
            np.arange(1, 10)*1000, len(x), replace=True)
          ).astype(str)
        )
      )

(fisd_dummy
  .to_sql(name="fisd",
          con=tidy_finance,
          if_exists="replace",
          index=False)
)
```

## C.2.2  Dummy `trace_enhanced` data

Finally, we create a dummy bond transaction data for the fictional CUSIPs of the dummy
`fisd` data. We take the date range that we also analyze in the book and ensure that we
have at least five transactions per day to fulfill a filtering step in the book.

```
number_of_bonds = 100
start_date = pd.Timestamp("2014-01-01")
end_date = pd.Timestamp("2016-11-30")

bonds_panel = pd.DataFrame({
  "cusip_id": np.tile(
    fisd_dummy["complete_cusip"],
    (end_date-start_date).days+1
  ),
  "trd_exctn_dt": np.repeat(
```

```
      pd.date_range(start_date, end_date), len(fisd_dummy)
  )
})

trace_enhanced_dummy = (pd.concat([bonds_panel]*5)
  .assign(
    trd_exctn_tm = lambda x: pd.to_datetime(
      x["trd_exctn_dt"].astype(str)+" " +
      np.random.randint(0, 24, size=len(x)).astype(str)+":" +
      np.random.randint(0, 60, size=len(x)).astype(str)+":" +
      np.random.randint(0, 60, size=len(x)).astype(str)
    ),
    rptd_pr=np.random.uniform(10, 200, len(bonds_panel)*5),
    entrd_vol_qt=1000*np.random.choice(
      range(1,21), len(bonds_panel)*5, replace=True
    ),
    yld_pt=np.random.uniform(-10, 10, len(bonds_panel)*5),
    rpt_side_cd=np.random.choice(
      ["B", "S"], len(bonds_panel)*5, replace=True
    ),
    cntra_mp_id=np.random.choice(
      ["C", "D"], len(bonds_panel)*5, replace=True
    )
  )
  .reset_index(drop=True)
)

(trace_enhanced_dummy
  .to_sql(name="trace_enhanced",
          con=tidy_finance,
          if_exists="replace",
          index=False)
)
```

As stated in the introduction, the data does *not* contain any samples of the original data. We merely generate random numbers for all columns of the tables that we use throughout this book.

# D

---

# *Clean Enhanced TRACE with Python*

This appendix contains code to clean enhanced TRACE with Python. It is also available via the following GitHub gist[1]. Hence, you could also source the file with the following chunk.

```
gist_url = (
  "https://gist.githubusercontent.com/patrick-weiss/"
  "86ddef6de978fbdfb22609a7840b5d8b/raw/"
  "8fbcc6c6f40f537cd3cd37368be4487d73569c6b/"
)


with httpimport.remote_repo(gist_url):
  from clean_enhanced_TRACE_python import clean_enhanced_trace
```

We need this function in Chapter 5 to download and clean enhanced TRACE trade messages following Dick-Nielsen (2009) and Dick-Nielsen (2014) for enhanced TRACE specifically. This code is based on the resources provided by the project Open Source Bond Asset Pricing[2] and their related publication Dickerson et al. (2023). We encourage that you acknowledge their effort. Relatedly, WRDS provides SAS code to clean enhanced TRACE data.

The function takes a vector of CUSIPs (in `cusips`), a connection to WRDS (`connection`) explained in Chapter 3, and a start and end date (`start_date` and `end_date`, respectively). Specifying too many CUSIPs will result in very slow downloads and a potential failure due to the size of the request to WRDS. The dates should be within the coverage of TRACE itself, i.e., starting after 2002, and the dates should be supplied as a string indicating MM/DD/YYYY. The output of the function contains all valid trade messages for the selected CUSIPs over the specified period.

```
def clean_enhanced_trace(cusips,
                         connection,
                         start_date="'01/01/2002'",
                         end_date="'12/31/2023'"):
  """Clean enhanced TRACE data."""

  import pandas as pd
  import numpy as np

  # Load main file
  trace_query = (
    "SELECT cusip_id, bond_sym_id, trd_exctn_dt, "
        "trd_exctn_tm, days_to_sttl_ct, lckd_in_ind, "
        "wis_fl, sale_cndtn_cd, msg_seq_nb, "
```

---

[1]https://gist.githubusercontent.com/patrick-weiss/86ddef6de978fbdfb22609a7840b5d8b
[2]https://openbondassetpricing.com/

```
            "trc_st, trd_rpt_dt, trd_rpt_tm, "
            "entrd_vol_qt, rptd_pr, yld_pt, "
            "asof_cd, orig_msg_seq_nb, rpt_side_cd, "
            "cntra_mp_id, stlmnt_dt, spcl_trd_fl "
    "FROM trace.trace_enhanced "
  f"WHERE cusip_id IN {cusips} "
        f"AND trd_exctn_dt BETWEEN {start_date} AND {end_date},"
)


trace_all = pd.read_sql_query(
  sql=trace_query,
  con=connection,
  parse_dates={"trd_exctn_dt","trd_rpt_dt", "stlmnt_dt"}
)


# Post 2012-06-02
## Trades (trc_st = T) and correction (trc_st = R)
trace_post_TR = (trace_all
  .query("trc_st in ['T', 'R']")
  .query("trd_rpt_dt >= '2012-06-02'")
)


# Cancellations (trc_st = X) and correction cancellations (trc_st = C)
trace_post_XC = (trace_all
  .query("trc_st in ['X', 'C']")
  .query("trd_rpt_dt >= '2012-06-02'")
  .get(["cusip_id", "msg_seq_nb", "entrd_vol_qt",
        "rptd_pr", "rpt_side_cd", "cntra_mp_id",
        "trd_exctn_dt", "trd_exctn_tm"])
  .assign(drop=True)
)


## Cleaning corrected and cancelled trades
trace_post_TR = (trace_post_TR
  .merge(trace_post_XC, how="left")
  .query("drop != True")
  .drop(columns="drop")
)


# Reversals (trc_st = Y)
trace_post_Y = (trace_all
  .query("trc_st == 'Y'")
  .query("trd_rpt_dt >= '2012-06-02'")
  .get(["cusip_id", "orig_msg_seq_nb", "entrd_vol_qt",
        "rptd_pr", "rpt_side_cd", "cntra_mp_id",
        "trd_exctn_dt", "trd_exctn_tm"])
  .assign(drop=True)
  .rename(columns={"orig_msg_seq_nb": "msg_seq_nb"})
)
```

```python
# Clean reversals
## Match the orig_msg_seq_nb of Y-message to msg_seq_nb of main message
trace_post = (trace_post_TR
  .merge(trace_post_Y, how="left")
  .query("drop != True")
  .drop(columns="drop")
)


# Pre 06-02-12
## Trades (trc_st = T)
trace_pre_T = (trace_all
  .query("trd_rpt_dt < '2012-06-02'")
)


# Cancellations (trc_st = C)
trace_pre_C = (trace_all
  .query("trc_st == 'C'")
  .query("trd_rpt_dt < '2012-06-02'")
  .get(["cusip_id", "orig_msg_seq_nb", "entrd_vol_qt",
        "rptd_pr", "rpt_side_cd", "cntra_mp_id",
        "trd_exctn_dt", "trd_exctn_tm"])
  .assign(drop=True)
  .rename(columns={"orig_msg_seq_nb": "msg_seq_nb"})
)


# Remove cancellations from trades
## Match orig_msg_seq_nb of C-message to msg_seq_nb of main message
trace_pre_T = (trace_pre_T
  .merge(trace_pre_C, how="left")
  .query("drop != True")
  .drop(columns="drop")
)


# Corrections (trc_st = W)
trace_pre_W = (trace_all
  .query("trc_st == 'W'")
  .query("trd_rpt_dt < '2012-06-02'")
)


# Implement corrections in a loop
## Correction control
correction_control = len(trace_pre_W)
correction_control_last = len(trace_pre_W)

## Correction loop
while (correction_control > 0):
  # Create placeholder
  ## Only identifying columns of trace_pre_T (for joins)
  placeholder_trace_pre_T = (trace_pre_T
    .get(["cusip_id", "trd_exctn_dt", "msg_seq_nb"])
```

```python
    .rename(columns={"msg_seq_nb": "orig_msg_seq_nb"})
    .assign(matched_T=True)
)


# Corrections that correct some msg
trace_pre_W_correcting = (trace_pre_W
    .merge(placeholder_trace_pre_T, how="left")
    .query("matched_T == True")
    .drop(columns="matched_T")
)


# Corrections that do not correct some msg
trace_pre_W = (trace_pre_W
    .merge(placeholder_trace_pre_T, how="left")
    .query("matched_T != True")
    .drop(columns="matched_T")
)


# Create placeholder
## Only identifying columns of trace_pre_W_correcting (for anti-joins)
placeholder_trace_pre_W_correcting = (trace_pre_W_correcting
    .get(["cusip_id", "trd_exctn_dt", "orig_msg_seq_nb"])
    .rename(columns={"orig_msg_seq_nb": "msg_seq_nb"})
    .assign(corrected=True)
)


# Delete msgs that are corrected
trace_pre_T = (trace_pre_T
    .merge(placeholder_trace_pre_W_correcting, how="left")
    .query("corrected != True")
    .drop(columns="corrected")
)


# Add correction msgs
trace_pre_T = pd.concat([trace_pre_T, trace_pre_W_correcting])


# Escape if no corrections remain or they cannot be matched
correction_control = len(trace_pre_W)


if correction_control == correction_control_last:
    break
else:
    correction_control_last = len(trace_pre_W)
    continue

# Reversals (asof_cd = R)
## Record reversals
trace_pre_R = (trace_pre_T
    .query("asof_cd == 'R'")
    .sort_values(["cusip_id", "trd_exctn_dt",
```

```
                    "trd_exctn_tm", "trd_rpt_dt", "trd_rpt_tm"])
)

## Prepare final data
trace_pre = (trace_pre_T
  .query(
    "asof_cd == None | asof_cd.isnull() | asof_cd not in ['R', 'X', 'D']"
  )
  .sort_values(["cusip_id", "trd_exctn_dt",
                "trd_exctn_tm", "trd_rpt_dt", "trd_rpt_tm"])
)


## Add grouped row numbers
trace_pre_R["seq"] = (trace_pre_R
  .groupby(["cusip_id", "trd_exctn_dt", "entrd_vol_qt",
            "rptd_pr", "rpt_side_cd", "cntra_mp_id"])
  .cumcount()
)

trace_pre["seq"] = (trace_pre
  .groupby(["cusip_id", "trd_exctn_dt", "entrd_vol_qt",
            "rptd_pr", "rpt_side_cd", "cntra_mp_id"])
  .cumcount()
)


## Select columns for reversal cleaning
trace_pre_R = (trace_pre_R
  .get(["cusip_id", "trd_exctn_dt", "entrd_vol_qt",
       "rptd_pr", "rpt_side_cd", "cntra_mp_id", "seq"])
  .assign(reversal=True)
)


## Remove reversals and the reversed trade
trace_pre = (trace_pre
  .merge(trace_pre_R, how="left")
  .query("reversal != True")
  .drop(columns=["reversal", "seq"])
)


# Combine pre and post trades
trace_clean = pd.concat([trace_pre, trace_post])

# Keep agency sells and unmatched agency buys
trace_agency_sells = (trace_clean
  .query("cntra_mp_id == 'D' & rpt_side_cd == 'S'")
)

# Placeholder for trace_agency_sells with relevant columns
placeholder_trace_agency_sells = (trace_agency_sells
  .get(["cusip_id", "trd_exctn_dt",
```

```python
                "entrd_vol_qt", "rptd_pr"])
    .assign(matched=True)
)


# Agency buys that are unmatched
trace_agency_buys_filtered = (trace_clean
  .query("cntra_mp_id == 'D' & rpt_side_cd == 'B'")
  .merge(placeholder_trace_agency_sells, how="left")
  .query("matched != True")
  .drop(columns="matched")
)


# Non-agency
trace_nonagency = (trace_clean
  .query("cntra_mp_id == 'C'")
)


# Agency cleaned
trace_clean = pd.concat([trace_nonagency,
                         trace_agency_sells,
                         trace_agency_buys_filtered])

# Additional Filters
trace_add_filters = (trace_clean
  .assign(
    days_to_sttl_ct2 = lambda x: (
      (x["stlmnt_dt"]-x["trd_exctn_dt"]).dt.days
    )
  )
  .assign(
    days_to_sttl_ct = lambda x: pd.to_numeric(
      x["days_to_sttl_ct"], errors='coerce'
    )
  )
  .query("days_to_sttl_ct.isnull() | days_to_sttl_ct <= 7")
  .query("days_to_sttl_ct2.isnull() | days_to_sttl_ct2 <= 7")
  .query("wis_fl == 'N'")
  .query("spcl_trd_fl.isnull() | spcl_trd_fl == ''")
  .query("asof_cd.isnull() | asof_cd == ''")
)


# Only keep necessary columns
trace_final = (trace_add_filters
  .sort_values(["cusip_id", "trd_exctn_dt", "trd_exctn_tm"])
  .get(["cusip_id", "trd_exctn_dt", "trd_exctn_tm", "rptd_pr",
        "entrd_vol_qt", "yld_pt", "rpt_side_cd", "cntra_mp_id"])
)


return trace_final
```

# E

## Cover Image

The cover of the book is inspired by the fast growing generative art community in R. Generative art refers to art that in whole or in part has been created with the use of an autonomous system. Instead of creating random dynamics, we rely on what is core to the book: The evolution of financial markets. Each circle corresponds to one of the twelve Fama-French industry portfolios, whereas each bar represents the average annual return between 1927 and 2022. The bar color is determined by the standard deviation of returns for each industry. The few lines of code below replicate the entire figure.

```python
import pandas as pd
import numpy as np
import matplotlib.pyplot as plt
import pandas_datareader as pdr

from datetime import datetime
from matplotlib.colors import LinearSegmentedColormap

main_colors = ["#3B9AB2", "#78B7C5", "#EBCC2A", "#E1AF00", "#F21A00"]
colormap = LinearSegmentedColormap.from_list("custom_colormap", main_colors)

industries_ff_daily_raw = pdr.DataReader(
  name="12_Industry_Portfolios_daily",
  data_source="famafrench",
  start="1927-01-01",
  end="2022-12-31")[0]

industries_ff_daily = (industries_ff_daily_raw
  .divide(100)
  .reset_index(names="date")
  .assign(date=lambda x: pd.to_datetime(x["date"].astype(str)))
  .rename(str.lower, axis="columns")
)

industries_long = (industries_ff_daily
  .melt(id_vars="date", var_name="name", value_name="value")
)

industries_order = sorted(industries_long["name"].unique())

data_plot = (industries_long
  .assign(year=industries_long["date"].dt.to_period("Y"))
  .groupby(["year", "name"])
```

```
    .aggregate(total=("value", "mean"),
               vola=("value", "std"))
    .reset_index()
    .assign(
      vola_ntile=lambda x: pd.qcut(x["vola"], 42, labels=False)
    )
)


dpi = 300
width = 2400/dpi
height = 1800/dpi
num_cols = 4
num_rows = int(len(industries_order)/num_cols)
fig, axs = plt.subplots(
  num_rows, num_cols,
  constrained_layout=True,
  subplot_kw={"projection": "polar"},
  figsize=(width, height),
  dpi=dpi
)
axs = axs.flatten()

for i in enumerate(industries_order):

    df = data_plot.copy().query(f'name == "{i[1]}"')
    min_value = df["total"].min()
    max_value = df["total"].max()
    std_value = df["total"].std()
    df["total"] = 2*(df["total"]-min_value)/(max_value-min_value)-1

    angles = np.linspace(0, 2*np.pi, len(df), endpoint=False)
    values = df["total"].values
    width = 2*np.pi/len(values)
    offset = np.pi/2

    ax = axs[i[0]]
    ax.set_theta_offset(offset)
    ax.set_ylim(-std_value*1400, 1)
    ax.set_frame_on(False)
    ax.xaxis.grid(False)
    ax.yaxis.grid(False)
    ax.set_xticks([])
    ax.set_yticks([])

    color_values = df["vola_ntile"].values
    normalize = plt.Normalize(min(color_values), max(color_values))
    colors = colormap(normalize(color_values))

    ax.bar(
      angles, values,
```

```
        width=width, color=colors, edgecolor="white", linewidth=0.2
    )

plt.tight_layout()
plt.subplots_adjust(wspace=-0.2, hspace=-0.1)
plt.gcf().savefig(
  "images/cover-image.png", dpi = 300, pad_inches=0, transparent=False
)
```

# Bibliography

Abadie, A., Athey, S., Imbens, G. W., and Wooldridge, J. (2017). When should you adjust standard errors for clustering? *Working Paper*.

Allaire, J., Teague, C., Scheidegger, C., Xie, Y., and Dervieux, C. (2023). Quarto. https://github.com/quarto-dev/quarto-cli. Version 1.4.

Arnott, R., Harvey, C. R., and Markowitz, H. (2019). A backtesting protocol in the era of machine learning. *The Journal of Financial Data Science*, 1(1):64–74.

Aroussi, R. (2023). yfinance. https://pypi.org/project/yfinance/. Version 0.2.31.

Avramov, D., Cheng, S., and Metzker, L. (2022). Machine learning vs. economic restrictions: Evidence from stock return predictability. *Management Science*, 69(5):2547–3155.

Avramov, D., Cheng, S., Metzker, L., and Voigt, S. (2023). Integrating factor models. *The Journal of Finance*, 78(3):1593–1646.

Bai, J., Bali, T. G., and Wen, Q. (2019). Common risk factors in the cross-section of corporate bond returns. *Journal of Financial Economics*, 131(3):619–642.

Balduzzi, P. and Lynch, A. W. (1999). Transaction costs and predictability: Some utility cost calculations. *Journal of Financial Economics*, 52(1):47–78.

Balduzzi, P. and Lynch, A. W. (2000). Predictability and transaction costs: The impact on rebalancing rules and behavior. *The Journal of Finance*, 55(5):2285–2309.

Bali, T. G., Engle, R. F., and Murray, S. (2016). *Empirical asset pricing: The cross section of stock returns*. John Wiley & Sons.

Ball, R. (1978). Anomalies in relationships between securities' yields and yield-surrogates. *Journal of Financial Economics*, 6(2–3):103–126.

Banz, R. W. (1981). The relationship between return and market value of common stocks. *Journal of Financial Economics*, 9(1):3–18.

Bayer, M. (2012). Sqlalchemy. In Brown, A. and Wilson, G., editors, *The Architecture of Open Source Applications Volume II: Structure, Scale, and a Few More Fearless Hacks*. aosabook.org.

Bessembinder, H., Kahle, K. M., Maxwell, W. F., and Xu, D. (2008). Measuring abnormal bond performance. *Review of Financial Studies*, 22(10):4219–4258.

Bessembinder, H., Maxwell, W., and Venkataraman, K. (2006). Market transparency, liquidity externalities, and institutional trading costs in corporate bonds. *Journal of Financial Economics*, 82(2):251–288.

Black, F. and Scholes, M. (1973). The pricing of options and corporate liabilities. *Journal of Political Economy*, 81(3):637–654.

Brandt, M. W. (2010). Portfolio choice problems. In Ait-Sahalia, Y. and Hansen, L. P., editors, *Handbook of Financial Econometrics: Tools and Techniques*, volume 1 of *Handbooks in Finance*, pages 269–336. North-Holland.

Brandt, M. W., Santa-Clara, P., and Valkanov, R. (2009). Parametric portfolio policies: Exploiting characteristics in the cross-section of equity returns. *Review of Financial Studies*, 22(9):3411–3447.

Brown, S. J. (1976). *Optimal portfolio choice under uncertainty: A Bayesian approach*. Phd thesis, University of Chicago.

Bryan, J. (2022). Happy Git and GitHub for the useR. https://github.com/jennybc/happy-git-with-r.

Bryzgalova, S., Pelger, M., and Zhu, J. (2022). Forest through the trees: Building cross-sections of stock returns. *Working Paper*.

Cameron, A. C., Gelbach, J. B., and Miller, D. L. (2011). Robust inference with multiway clustering. *Journal of Business & Economic Statistics*, 29(2):238–249.

Campbell, J. Y. (1987). Stock returns and the term structure. *Journal of Financial Economics*, 18(2):373–399.

Campbell, J. Y., Hilscher, J., and Szilagyi, J. (2008). In search of distress risk. *The Journal of Finance*, 63(6):2899–2939.

Campbell, J. Y., Lo, A. W., MacKinlay, A. C., and Whitelaw, R. F. (1998). The econometrics of financial markets. *Macroeconomic Dynamics*, 2(4):559–562.

Campbell, J. Y. and Shiller, R. J. (1988). Stock prices, earnings, and expected dividends. *The Journal of Finance*, 43(3):661–676.

Campbell, J. Y. and Vuolteenaho, T. (2004). Inflation illusion and stock prices. *American Economic Review*, 94(2):19–23.

Campbell, J. Y. and Yogo, M. (2006). Efficient tests of stock return predictability. *Journal of Financial Economics*, 81(1):27–60.

Chen, A. Y. and Zimmermann, T. (2022). Open source cross-sectional asset pricing. *Critical Finance Review*, 11(2):207–264.

Chen, H.-Y., Lee, A. C., and Lee, C.-F. (2015). Alternative errors-in-variables models and their applications in finance research. *The Quarterly Review of Economics and Finance*, 58:213–227.

Chen, L., Pelger, M., and Zhu, J. (2023). Deep learning in asset pricing. *Management Science*.

Chopra, V. K. and Ziemba, W. T. (1993). The effect of errors in means, variances, and covariances on optimal portfolio choice. *Journal of Portfolio Management*, 19(2):6–11.

Cochrane, J. H. (2005). Writing tips for PhD students. *Note*.

Cochrane, J. H. (2009). *Asset pricing (revised edition)*. Princeton University Press.

Cochrane, J. H. (2011). Presidential address: Discount rates. *The Journal of Finance*, 66(4):1047–1108.

Coqueret, G. and Guida, T. (2020). *Machine learning for factor investing: R version*. Chapman and Hall/CRC.

Coqueret, G. and Guida, T. (2023). *Machine learning for factor investing: Python version.* Chapman and Hall/CRC.

Davis, M. H. A. and Norman, A. R. (1990). Portfolio selection with transaction costs. *Mathematics of Operations Research*, 15(4):676–713.

De Prado, M. L. (2018). *Advances in financial machine learning.* John Wiley & Sons.

DeMiguel, V., Martín-Utrera, A., and Nogales, F. J. (2015). Parameter uncertainty in multiperiod portfolio optimization with transaction costs. *Journal of Financial and Quantitative Analysis*, 50(6):1443–1471.

DeMiguel, V., Nogales, F. J., and Uppal, R. (2014). Stock return serial dependence and out-of-sample portfolio performance. *Review of Financial Studies*, 27(4):1031–1073.

Dick-Nielsen, J. (2009). Liquidity biases in TRACE. *The Journal of Fixed Income*, 19(2):43–55.

Dick-Nielsen, J. (2014). How to clean enhanced TRACE data. *Working Paper*.

Dick-Nielsen, J., Feldhütter, P., and Lando, D. (2012). Corporate bond liquidity before and after the onset of the subprime crisis. *Journal of Financial Economics*, 103(3):471–492.

Dickerson, A., Mueller, P., and Robotti, C. (2023). Priced risk in corporate bonds. *Journal of Financial Economics*, 150(2):103707.

Dixon, M. F., Halperin, I., and Bilokon, P. (2020). *Machine learning in finance.* Springer.

Donald, S. and Lang, K. (2007). Inference with difference-in-differences and other panel data. *The Review of Economics and Statistics*, 89(2):221–233.

Easley, D., de Prado, M., O'Hara, M., and Zhang, Z. (2020). Microstructure in the machine age. *Review of Financial Studies*, 34(7):3316–3363.

Edwards, A. K., Harris, L. E., and Piwowar, M. S. (2007). Corporate bond market transaction costs and transparency. *The Journal of Finance*, 62(3):1421–1451.

Erickson, T. and Whited, T. M. (2012). Treating measurement error in Tobin's q. *Review of Financial Studies*, 25(4):1286–1329.

Fama, E. F. and French, K. R. (1989). Business conditions and expected returns on stocks and bonds. *Journal of Financial Economics*, 25(1):23–49.

Fama, E. F. and French, K. R. (1992). The cross-section of expected stock returns. *The Journal of Finance*, 47(2):427–465.

Fama, E. F. and French, K. R. (1993). Common risk factors in the returns on stocks and bonds. *Journal of Financial Economics*, 33(1):3–56.

Fama, E. F. and French, K. R. (1997). Industry costs of equity. *Journal of Financial Economics*, 43(2):153–193.

Fama, E. F. and French, K. R. (2015). A five-factor asset pricing model. *Journal of Financial Economics*, 116(1):1–22.

Fama, E. F. and MacBeth, J. D. (1973). Risk, return, and equilibrium: Empirical tests. *Journal of Political Economy*, 81(3):607–636.

Fan, J., Fan, Y., and Lv, J. (2008). High dimensional covariance matrix estimation using a factor model. *Journal of Econometrics*, 147(1):186–197.

Fazzari, S. M., Hubbard, R. G., Petersen, B. C., Blinder, A. S., and Poterba, J. M. (1988). Financing constraints and corporate investment. *Brookings Papers on Economic Activity*, 1988(1):141–206.

Foundation, P. S. and JetBrains (2022). Python developers survey 2022 results. https://lp.jetbrains.com/python-developers-survey-2022/.

Frazzini, A. and Pedersen, L. H. (2014). Betting against beta. *Journal of Financial Economics*, 111(1):1–25.

Gagliardini, P., Ossola, E., and Scaillet, O. (2016). Time-varying risk premium in large cross-sectional equity data sets. *Econometrica*, 84(3):985–1046.

Gareth, J., Daniela, W., Trevor, H., and Robert, T. (2013). *An introduction to statistical learning: With applications in R*. Springer.

Garlappi, L., Uppal, R., and Wang, T. (2007). Portfolio selection with parameter and model uncertainty: A multi-prior approach. *Review of Financial Studies*, 20(1):41–81.

Gârleanu, N. and Pedersen, L. H. (2013). Dynamic trading with predictable returns and transaction costs. *The Journal of Finance*, 68(6):2309–2340.

Gehrig, T., Sögner, L., and Westerkamp, A. (2020). Making portfolio policies work. *Working Paper*.

Goldstein, I., Koijen, R. S. J., and Mueller, H. M. (2021). COVID-19 and its impact on financial markets and the real economy. *Review of Financial Studies*, 34(11):5135–5148.

Gu, S., Kelly, B., and Xiu, D. (2020). Empirical asset pricing via machine learning. *Review of Financial Studies*, 33(5):2223–2273.

Gulen, H. and Ion, M. (2015). Policy uncertainty and corporate investment. *Review of Financial Studies*, 29(3):523–564.

Guo, H. (2006). On the out-of-sample predictability of stock market returns. *The Journal of Business*, 79(2):645–670.

Halling, M., Yu, J., and Zechner, J. (2021). Primary corporate bond markets and social responsibility. *Working Paper*.

Handler, L., Jankowitsch, R., and Pasler, A. (2022). The effects of ESG performance and preferences on us corporate bond prices. *Working Paper*.

Handler, L., Jankowitsch, R., and Weiss, P. (2021). Covenant prices of US corporate bonds. *Working Paper*.

Harris, C. R., Millman, K. J., van der Walt, S. J., Gommers, R., Virtanen, P., Cournapeau, D., Wieser, E., Taylor, J., Berg, S., Smith, N. J., Kern, R., Picus, M., Hoyer, S., van Kerkwijk, M. H., Brett, M., Haldane, A., del Río, J. F., Wiebe, M., Peterson, P., Gérard-Marchant, P., Sheppard, K., Reddy, T., Weckesser, W., Abbasi, H., Gohlke, C., and Oliphant, T. E. (2020). Array programming with NumPy. *Nature*, 585(7825):357–362.

Harvey, C. R. (2017). Presidential address: The scientific outlook in financial economics. *The Journal of Finance*, 72(4):1399–1440.

Harvey, C. R., Liu, Y., and Zhu, H. (2016). ... and the cross-section of expected returns. *Review of Financial Studies*, 29(1):5–68.

Hasler, M. (2021). Is the value premium smaller than we thought? *Working Paper*.

Hastie, T., Tibshirani, R., and Friedman, J. (2009). *The elements of statistical learning: Data mining, inference and prediction.* Springer, 2 edition.

Hautsch, N., Kyj, L. M., and Malec, P. (2015). Do high-frequency data improve high-dimensional portfolio allocations? *Journal of Applied Econometrics*, 30(2):263–290.

Hautsch, N. and Voigt, S. (2019). Large-scale portfolio allocation under transaction costs and model uncertainty. *Journal of Econometrics*, 212(1):221–240.

Hilpisch, Y. (2018). *Python for Finance.* O'Reilly Media, Inc., second edition.

Hoerl, A. E. and Kennard, R. W. (1970). Ridge regression: Applications to nonorthogonal problems. *Technometrics*, 12(1):69–82.

Hornik, K. (1991). Approximation capabilities of multilayer feedforward networks. *Neural Networks*, 4(2):251–257.

Hou, K., Xue, C., and Zhang, L. (2014). Digesting anomalies: An investment approach. *Review of Financial Studies*, 28(3):650–705.

Hou, K., Xue, C., and Zhang, L. (2020). Replicating anomalies. *Review of Financial Studies*, 33(5):2019–2133.

Huang, J.-Z. and Shi, Z. (2021). What do we know about corporate bond returns? *Annual Review of Financial Economics*, 13(1):363–399.

Hull, J. C. (2020). *Machine learning in business. An introduction to the world of data science.* Independently published.

Huynh, T. D. and Xia, Y. (2021). Climate change news risk and corporate bond returns. *Journal of Financial and Quantitative Analysis*, 56(6):1985–2009.

Jacobsen, B. (2014). Some research and writing tips. *Note.*

Jagannathan, R. and Ma, T. (2003). Risk reduction in large portfolios: Why imposing the wrong constraints helps. *The Journal of Finance*, 58(4):1651–1684.

Jegadeesh, N. and Titman, S. (1993). Returns to buying winners and selling losers: Implications for stock market efficiency. *The Journal of Finance*, 48(1):65–91.

Jensen, T. I., Kelly, B. T., Malamud, S., and Pedersen, L. H. (2022). Machine learning and the implementable efficient frontier. *Working Paper.*

Jiang, J., Kelly, B. T., and Xiu, D. (2023). (Re-)Imag(in)ing Price Trends. *The Journal of Finance*, 78(6):3193–3249.

Jobson, D. J. and Korkie, B. (1980). Estimation for Markowitz efficient portfolios. *Journal of the American Statistical Association*, 75(371):544–554.

Jorion, P. (1986). Bayes-Stein estimation for portfolio analysis. *Journal of Financial and Quantitative Analysis*, 21(03):279–292.

Kan, R. and Zhou, G. (2007). Optimal portfolio choice with parameter uncertainty. *Journal of Financial and Quantitative Analysis*, 42(3):621–656.

Kelliher, C. (2022). *Quantitative Finance with Python.* Chapman and Hall/CRC, 1st edition.

Kelly, B. T., Palhares, D., and Pruitt, S. (2021). Modeling corporate bond returns. *Working Paper.*

Kibirige, H. (2023a). mizani: A port of the scales package for use with plotnine. https://pypi.org/project/mizani/. Version 0.9.3.

Kibirige, H. (2023b). plotnine: An implementation of the grammar of graphics in python. https://pypi.org/project/plotnine/. Version 0.12.3.

Kiesling, L. (2003). Writing tips for economics (and pretty much anything else). *Note*.

Kim, D. (1995). The errors in the variables problem in the cross-section of expected stock returns. *The Journal of Finance*, 50(5):1605–1634.

Kothari, S. P. and Shanken, J. A. (1997). Book-to-market, dividend yield, and expected market returns: A time-series analysis. *Journal of Financial Economics*, 44(2):169–203.

Kraft, D. (1994). Algorithm 733: TOMP–Fortran Modules for Optimal Control Calculations. *ACM Trans. Math. Softw.*, 20(3):262–281.

Lamont, O. (1998). Earnings and expected returns. *The Journal of Finance*, 53(5):1563–1587.

Ledoit, O. and Wolf, M. (2003). Improved estimation of the covariance matrix of stock returns with an application to portfolio selection. *Journal of Empirical Finance*, 10(5):603–621.

Ledoit, O. and Wolf, M. (2004). Honey, I shrunk the sample covariance matrix. *The Journal of Portfolio Management*, 30(4):110–119.

Ledoit, O. and Wolf, M. (2012). Nonlinear shrinkage estimation of large-dimensional covariance matrices. *The Annals of Statistics*, 40(2):1024–1060.

Lintner, J. (1965). Security prices, risk, and maximal gains from diversification. *The Journal of Finance*, 20(4):587–615.

Magill, M. J. P. and Constantinides, G. M. (1976). Portfolio selection with transactions costs. *Journal of Economic Theory*, 13(2):245–263.

Markowitz, H. (1952). Portfolio selection. *The Journal of Finance*, 7(1):77–91.

Mclean, R. D. and Pontiff, J. (2016). Does academic research destroy stock return predictability? *The Journal of Finance*, 71(1):5–32.

Menkveld, A. J., Dreber, A., Holzmeister, F., Huber, J., Johannesson, M., Kirchler, M., Neusüss, S., Razen, M., and Weitzel, U. (2021). Non-standard errors. *Working Paper*.

Merton, R. C. (1972). An analytic derivation of the efficient portfolio frontier. *Journal of Financial and Quantitative Analysis*, 7(4):1851–1872.

Mossin, J. (1966). Equilibrium in a capital asset market. *Econometrica*, 34(4):768–783.

Mullainathan, S. and Spiess, J. (2017). Machine learning: An applied econometric approach. *Journal of Economic Perspectives*, 31(2):87–106.

Nagel, S. (2021). *Machine learning in asset pricing*. Princeton University Press.

Newey, W. K. and West, K. D. (1987). A simple, positive semi-definite, heteroskedasticity and autocorrelation consistent covariance Matrix. *Econometrica*, 55(3):703–708.

O'Hara, M. and Zhou, X. A. (2021). Anatomy of a liquidity crisis: Corporate bonds in the COVID-19 crisis. *Journal of Financial Economics*, 142(1):46–68.

Pedregosa, F., Varoquaux, G., Gramfort, A., Michel, V., Thirion, B., Grisel, O., Blondel, M., Prettenhofer, P., Weiss, R., Dubourg, V., Vanderplas, J., Passos, A., Cournapeau, D., Brucher, M., Perrot, M., and Duchesnay, E. (2011). Scikit-learn: Machine learning in Python. *Journal of Machine Learning Research*, 12:2825–2830.

Peters, R. H. and Taylor, L. A. (2017). Intangible capital and the investment-q relation. *Journal of Financial Economics*, 123(2):251–272.

Petersen, M. A. (2008). Estimating standard errors in finance panel data sets: Comparing approaches. *Review of Financial Studies*, 22(1):435–480.

Pflug, G., Pichler, A., and Wozabal, D. (2012). The 1/N investment strategy is optimal under high model ambiguity. *Journal of Banking & Finance*, 36(2):410–417.

Python Software Foundation (2023). Python language reference, version 3.10.11.

Roberts, M. R. and Whited, T. M. (2013). Endogeneity in empirical corporate finance. In *Handbook of the Economics of Finance*, volume 2, pages 493–572. Elsevier.

Scheuch, C., Voigt, S., and Weiss, P. (2023). *Tidy Finance with R*. Chapman and Hall/CRC, 1st edition.

Seabold, S. and Perktold, J. (2010). statsmodels: Econometric and statistical modeling with python. In *9th Python in Science Conference*.

Seltzer, L. H., Starks, L., and Zhu, Q. (2022). Climate regulatory risk and corporate bonds. *Working Paper*.

Shanken, J. (1992). On the estimation of beta-pricing models. *Review of Financial Studies*, 5(1):1–33.

Sharpe, W. F. (1964). Capital asset prices: A theory of market equilibrium under conditions of risk . *The Journal of Finance*, 19(3):425–442.

Sheppard, K. (2023). linearmodels: Instrumental variable and linear panel models for python. https://pypi.org/project/linearmodels/. Version 5.3.

Soebhag, A., Van Vliet, B., and Verwijmeren, P. (2022). Mind your sorts. *Working Paper*.

Team, J. D. (2023). Joblib: running python functions as pipeline jobs. https://joblib.readthedocs.io/. Version 1.3.2.

Tibshirani, R. (1996). Regression shrinkage and selection via the LASSO. *Journal of the Royal Statistical Society. Series B (Methodological)*, 58(1):267–288.

Torakis, J. (2023). httpimport: Module for remote in-memory python package/module loading through http. https://pypi.org/project/httpimport/. Version 1.3.1.

Tsay, R. S. (2010). *Analysis of financial time series*. John Wiley & Sons.

Virtanen, P., Gommers, R., Oliphant, T. E., Haberland, M., Reddy, T., Cournapeau, D., Burovski, E., Peterson, P., Weckesser, W., Bright, J., van der Walt, S. J., Brett, M., Wilson, J., Millman, K. J., Mayorov, N., Nelson, A. R. J., Jones, E., Kern, R., Larson, E., Carey, C. J., Polat, İ., Feng, Y., Moore, E. W., VanderPlas, J., Laxalde, D., Perktold, J., Cimrman, R., Henriksen, I., Quintero, E. A., Harris, C. R., Archibald, A. M., Ribeiro, A. H., Pedregosa, F., van Mulbregt, P., and SciPy 1.0 Contributors (2020). SciPy 1.0: Fundamental Algorithms for Scientific Computing in Python. *Nature Methods*, 17:261–272.

Walter, D., Weber, R., and Weiss, P. (2022). Non-standard errors in portfolio sorts. *Working Paper*.

Wang, Z. (2005). A shrinkage approach to model uncertainty and asset allocation. *Review of Financial Studies*, 18(2):673–705.

Wasserstein, R. L. and Lazar, N. A. (2016). The ASA Statement on p-Values: Context, process, and purpose. *The American Statistician*, 70(2):129–133.

Weiming, J. M. (2019). *Mastering Python for Finance*. Packt, 2nd edition.

Welch, I. and Goyal, A. (2008). A comprehensive look at the empirical performance of equity premium prediction. *Review of Financial Studies*, 21(4):1455–1508.

Wes McKinney (2010). Data Structures for Statistical Computing in Python. In Stéfan van der Walt and Jarrod Millman, editors, *Proceedings of the 9th Python in Science Conference*, pages 56–61.

Wickham, H. (2014). Tidy data. *Journal of Statistical Software*, 59(1):1–23.

Wickham, H. and Girlich, M. (2022). *tidyr: Tidy messy data*. R package version 1.2.1.

Wilkinson, L. (2012). *The grammar of graphics*. Springer.

Wooldridge, J. M. (2010). *Econometric analysis of cross section and panel data*. The MIT Press.

Zaffaroni, P. and Zhou, G. (2022). Asset pricing: Cross-section predictability. *Working Paper*.

Zou, H. and Hastie, T. (2005). Regularization and variable selection via the elastic net. *Journal of the Royal Statistical Society. Series B (Statistical Methodology)*, 67(2):301–320.

# Index

Printed in the United States
by Baker & Taylor Publisher Services